Reviews and critical articles covering the entire field of normal anatomy (cytology, histology, cyto- and histochemistry, electron microscopy, macroscopy, experimental morphology and embryology and comparative anatomy) are published in Advances in Anatomy, Embryology and Cell Biology. Papers dealing with anthropology and clinical morphology that aim to encourage cooperation between anatomy and related disciplines will also be accepted. Papers are normally commissioned. Original papers and communications may be submitted and will be considered for publication provided they meet the requirements of a review article and thus fit into the scope of "Advances". English language is preferred, but in exceptional cases French or German papers will be accepted.

It is a fundamental condition that submitted manuscripts have not been and will not simultaneously be submitted or published elsewhere. With the acceptance of a manuscript for publication, the publisher acquires full and exclusive copyright for all languages and countries.

Twenty-five copies of each paper are supplied free of charge.

Manuscripts should be addressed to

Prof. Dr. F. **BECK,** Howard Florey Institute, University of Melbourne, Parkville, 3000 Melbourne, Victoria, Australia

Prof. Dr. B. **CHRIST**, Anatomisches Institut der Universität Freiburg, Abteilung Anatomie II, Albertstr. 17, D-79104 Freiburg, Germany

Prof. Dr. W. **KRIZ,** Anatomisches Institut der Universität Heidelberg, Im Neuenheimer Feld 307, D-69120 Heidelberg, Germany

Prof. Dr. E. **MARANI**, Leiden University, Department of Physiology, Neuroregulation Group, P.O. Box 9604, 2300 RC Leiden, The Netherlands

Prof. Dr. R. **PUTZ**, Anatomische Anstalt der Universität München, Lehrstuhl Anatomie I, Pettenkoferstr. 11, D-80336 München, Germany

Prof. Dr. Dr. h.c. Y. **SANO,** Department of Anatomy, Kyoto Prefectural University of Medicine, Kawaramachi-Hirokoji, 602 Kyoto, Japan

Prof. Dr. Dr. h.c. T. H. **SCHIEBLER,** Anatomisches Institut der Universität, Koellikerstraße 6, D-97070 Würzburg, Germany

Prof. Dr. K. **ZILLES**, Universität Düsseldorf, Medizinische Einrichtungen, C. u. O. Vogt-Institut, Postfach 101007, D-40001 Düsseldorf, Germany

Advances in Anatomy
Embryology and Cell Biology

Vol. 156

Springer-Verlag Berlin Heidelberg GmbH

K. J. Barteczko M. I. Jacob

The Testicular Descent in Human

Origin, Development and Fate
of the Gubernaculum Hunteri,
Processus Vaginalis Peritonei,
and Gonadal Ligaments

With 60 Figures and 2 Tables

 Springer

Klaus J. Barteczko
Monika I. Jacob
Abteilung für Anatomie und Embryologie
Institut für Anatomie
Ruhr-Universität Bochum
Universitätsstraße 150
44780 Bochum, Germany
e-mail: monika.jacob@ruhr-uni-bochum.de
e-mail: klaus.barteczko@ruhr-uni-bochum.de

ISSN 0301-5556
ISBN 978-3-540-67315-6

Library of Congress-Cataloging-in-Publication-Data

Die Deutsche Bibliothek - CIP-Einheitsaufnahme
Barteczko, Klaus J.: The testicular descent in human : origin, development
and fate of the gubernaculum hunteri, processus vaginalis peritonei, and
gonadal ligaments ; with60 figures and 2 tables / K. J. Barteczko ; M. I. Jacob.
- Berlin ; Heidelberg ; New York : Springer, 2000
(Advances in anatomy, embryology, and cell biology, Vol. 156)
  ISBN 978-3-540-67315-6   ISBN 978-3-642-58353-7 (eBook)
  DOI 10.1007/978-3-642-58353-7

© Springer-Verlag Berlin Heidelberg 2000
Originally published by Springer-Verlag Berlin Heidelberg New York in 2000

Production: PRO EDIT GmbH, 69126 Heidelberg, Germany
Printed on acid-free paper - SPIN: 10718087   27/3136wg - 5 4 3 2 1 0

In memoriam to
Prof. Dr. Klaus V. Hinrichsen (1927–1997),
the great embryologist and teacher

Preface

Preparatory work for this monograph started in 1988 and 1989. The motive was provided by certain incongruities in the illustration of descensus testis for the chapter "Entwicklung der Genitalorgane" in the book *Humanembryologie*.

The evoked discussion was reason enough for our own research on the phenomenon and uncertainties of testicular descent in human. The basis of these investigations was the collection of human embryos founded by Prof. Hinrichsen in 1970 at the Department of Anatomy and Embryology, Ruhr-University Bochum.

The preparation of the material used in this paper has been done with the help of many staff members of our department to whom we feel obliged. We are especially indebted to Prof. Hinrichsen who prepared the foetuses and took the first steps of this project.

We wish also to acknowledge Dr. Heinz Jürgen Jacob for careful fixation and microdissection of specimens intended for scanning electron microscopy. We are also grateful to Mrs. Vera Mannheim for her involvement and her sensitivity in making scanning electron micrographs, and Mrs. Antje Jaeger for providing an expert and skilful photographic technique especially in the setting up of micrographs. Our thanks are also due to Mrs. Marion Köhn for skilful technical assistance and to all those technical assistants who worked in our department during the past 25 years and who were engaged in providing the serial sections.

The English text has been thoroughly revised by Prof. Dr. Holger Preuschoft, Bochum; his comments have been a great help to us.

Bochum, October 1999   KLAUS BARTECZKO and
MONIKA JACOB

Contents

1 Introduction

Why another review dealing with testicular descent? In 1996 Hutson et al. published a monograph in this series entitled Normal Testicular Descent and the Aetiology of Cryptorchidism. They stated: "Part of the dilemma is carried by lack of understanding of the normal process of descent, despite intensive research and study since the eighteenth century when Hunter and von Haller first described gubernaculum.... More needs to be known." Two hundred years of studies have brought many theories and contradictions.

Since experimental work is mainly performed on laboratory animals like rats and mice (e.g. Beasley and Hutson 1988; Bergh et al. 1978; Bjerklund Johansen 1988; Cain et al. 1995; Clarnette and Hutson 1997; Clarnette et al. 1996; Frey and Rajfer 1984; Frey et al. 1983; Goh et al. 1993; Griffiths 1993; Johansen et al. 1989; Kassim et al 1997; Lam et al. 1998; van der Schoot 1992; Shono et al 1994a,b, 1996, 1999; Spencer et al. 1991), dogs (Baumans et al. 1982, 1983, 1985), or pigs (Colenbrander et al. 1978; Fentener van Vlissingen et al. 1989; Heyns and de Klerk 1985; Heyns et al. 1986, 1989, 1990, 1993; Noordhuizen-Stassen et al. 1983), normal developmental processes during testicular descent in humans lies outside the scope of many scientists. Furthermore, the anatomy of the descent of the testis in the foetal calf has been reviewed by Hullinger and Wensing (1985). Yet, to avoid misleading conclusion drawn from analogies with laboratory animals, human material has to be studied. Species-specific differences are immense. For descent of the testis in the rat, Fujikake et al. (1989) might be consulted, and a comparison of this process in the rat with that in the pig is given by Wensing (1986).

Despite the great number of investigations in the past 100 years, in anatomy and embryology textbooks, the diagrams still are based on the model of Hertwig (1907) as seen in Patten (1968), Tuchmann-Duplessis and Haegel (1972), Sadler (1998) and others, showing the gubernaculum as an elastic band which extends from the gonad to the bottom of the scrotum. Since we did not see such a gubernaculum, this was reason and motivation enough to trace the situation back to that in human foetuses. We, therefore, studied structures involved in gonadal descent using histological sections, microdissection, immunohistochemistry and scanning electron microscopy. The latter method allows almost a three-dimensional view of the organs and ligaments involved in testicular descent. Equally, graphic reconstructions were made which help to understand the dimensions and relations of gonads and ligaments during this process.

It is impossible to discuss all the theories on testicular descent which have been reviewed by Backhouse (1964, 1982), Bramann (1884), Gier and Marion (1970), Habenicht and Neumann (1983), Hadziselimovic (1983), Hadziselimovic and Herzog

(1990), Heyns (1987), Heyns and Hutson (1995), Husmann and Levy (1995), Hutson and Beasley (1992), Hutson et al. (1990, 1994, 1996, 1997), Klaatsch (1890), Moszkowicz (1935, 1939), Weil (1884), Wensing (1988), Wensing and Colenbrander (1986), Youssef and Raslan (1971) and others. Thus, we will focus on review articles concentrating the wealth of data for the reader.

We also will focus on three main topics: 1, the inner descent; 2, origin, shape and fate of the gubernaculum Hunteri and the processus vaginalis peritonei; 3, origin of the ligaments of the ovary and uterus. Preliminary results have been published by Barteczko and Jacob (1999a), Barteczko et al. (1998), and Jacob et al. (1998).

2 Materials and Methods

This study is based on human embryos from the Collection of the Department of Anatomy and Embryology at the Ruhr University Bochum. The embryos and foetuses used in this study were fixed in formaldehyde or Bouin's fixative. The specimens were dehydrated, embedded in Parablast and serially sectioned with 8–10 μm thickness. They were alternatively stained by hematoxylin-eosin, Azan or trichrome Masson.

In older foetuses the abdomen was opened and the inguinal region and the scrotum were dissected.

Graphic reconstructions were made for the interesting parts of the series as already described by Barteczko and Jacob (1999b).

2.1
Scanning Electron Microscopy

Embryos were fixed by perfusion (3% glutaraldehyde in phosphate buffer 0.12 M, pH 7.3) through the umbilical vessels. After postfixation by immersion for 48 h in the same solution, the washing procedure was done with phosphate buffer with the addition of 8% glucose. After further fixation in 1% OsO4 (Dalton 1955), the specimens were dehydrated in methanol and critical point dried either with frigen or carbon dioxide. They were mounted with conducting silver and spattered with gold. A SEM SM 35 (Jeol) was used with a LAB6 Electron Gun-hexaboride cathode.

2.2
Immunohistochemistry

Formaldehyde or Bouin-fixed specimens, already stained and mounted, were used for immunohistochemistry after histological investigation. The coverslips were removed from the slides by xylene. The sections were rehydrated, washed in phosphate-buffered saline (PBS), and treated with 3% H2O2 in PBS for 5 min to inhibit endogenous peroxidase activity. After washing with three changes of PBS, the slides were incubated in normal goat serum diluted 1:5 in PBS for 20 min at room temperature in order to prevent subsequent non-specific absorption of antibodies. The monoclonal smooth muscle antibody (Dako M0851) was then applied at concentrations of 1:50 and 1:70 overnight at 4°C. The second antibody was a biotinylated goat anti-mouse antibody (Dako), followed by StreptABComplex/HRP (Dako). After DAB reaction with DAB-buffer tablets (Merck) for 10 min at room temperature, the brown product

3

of the enzymatic reaction appeared. Between each reaction washing in PBS with three changes was necessary. Finally, the sections were again dehydrated and mounted. For controls, the first antibody was omitted.

In the same manner, the proliferation marker anti-proliferating cell nuclear antigen(PCNA; Dako M0879) was employed using concentrations of 1:150 and 1:250.

4

3 Transabdominal (Inner) Descent of Gonads – Anlage of Diaphragm and Cranial Mesonephric Ligament

Testicular descent is usually described in two stages: the intra-abdominal passage and the passage through the abdominal wall into the scrotum (see the reviews of Heyns and Hutson 1995; Hutson et al. 1990). Yet, the inner descent has often been denied and some authors like Felix (1911) and Politzer and Zeitlhofer (1958) supposed that it does not really occur but becomes apparent due to degeneration of the cranial part of gonads and growth of the dorsal abdominal wall relative to the gubernaculum (Rajfer and Welsh 1977; van der Schoot 1993a).

The descending septum transversum or diaphragm and the intercalated so-called cranial gonadal ligament (cranial mesonephric ligaments) are crucial points in this process. Therefore, in the present report we re-examine the size and extension of this ligament in human embryos as well as its structure and blood vessels.

The proportional relations of identical notochordal segments in embryos at subsequent developing stages and length are compared. This reveals that no "explosion-like" increase of lengths in special vertebral segments exists which simulate an inner descend of gonads.

Special attention is paid to the level of origin (related to vertebral segments) of the testicular or ovarian arteries, as well as the inferior mesenteric artery and the renal arteries.

In the stage 14 embryo, 6.5 mm crown-rump length (CRL; Fig. 1a), the pleuroperitoneal membranes have not completed the septum transversum (Wells 1954). The cranial part of each mesonephros is situated in a pleuroperitoneal canal immediately dorsal to the paired pleuroperitoneal membrane. During closure of the canals, each mesonephros is connected with the septum transversum or anlage of the diaphragm by a well-developed ligament (Fig. 1b,c). This ligament always has an insertion near the forming ostium abdominale of the müllerian duct in a region where hydatides will eventually form. Also in embryos of 20 and 21 mm CRL whose enlarged gonad anlage reaches the upper pole of the mesonephros, no insertion of the ligament at the gonad was seen (Figs. 2, 3; see also Jirasek 1983, Figs. 116 and 118 described as cranial portion of the genital ridge). No cranial gonadal ligament exists; it should be named cranial mesonephric ligament. This situation does not change in older embryos (Fig. 4a), and in 50-mm CRL female embryos (Fig. 4b) a very thin ligament inserts into the mesonephros near its border to the gonad. This ligament does not contain any vessel for the blood supply of the gonads. In Fig. 5, the relationships of this ligament with the anlage of the diaphragm, gonad and regressing mesonephros are shown.

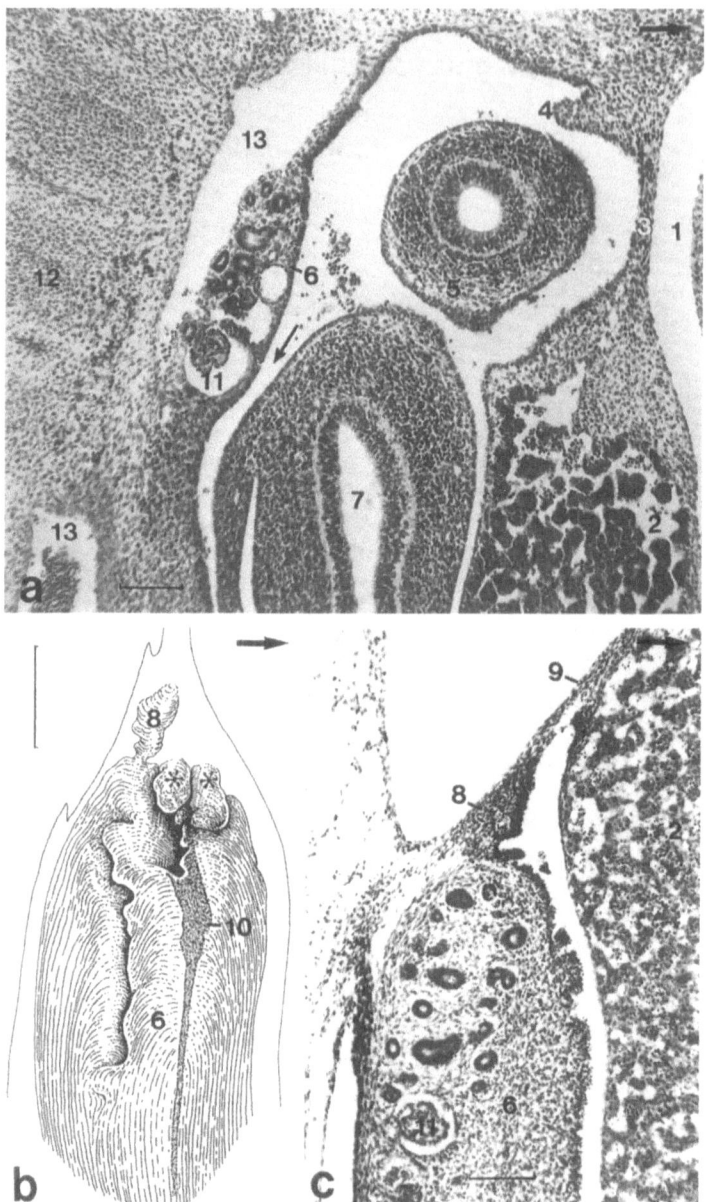

Fig. 1a-b. a Embryo of 6.5 mm CRL (crown-rump length). Sagittal section through the pleuroperitoneal canal. Medial view, left side. **b** Embryo of 16 mm CRL. Detailed reconstruction of the right mesonephros with the cranial mesonephric ligament and the funnel field where the abdominal ostium of the müllerian duct will develop. **c** Same embryo as in **b**. Sagittal section through the cranial part of the right mesonephros. *1*, pericardial cavity; *2*, liver; *3*, septum transversum; *4*, pleuroperitoneal membrane; *5*, lung bud; *6*, mesonephros; *7*, stomach; *8*, cranial mesonephric ligament; *9*, anlage of diaphragm; *10*, müllerian duct; *11*, mesonephric glomerulus; *12*, dorsal body wall; *13*, aorta. The *small arrow* indicates the pleuroperitoneal canal, the *large arrow* always shows ventral direction, *asterisks* indicate the dorsal and ventral lip in the "funnel field". *Bars*, 0.1 mm

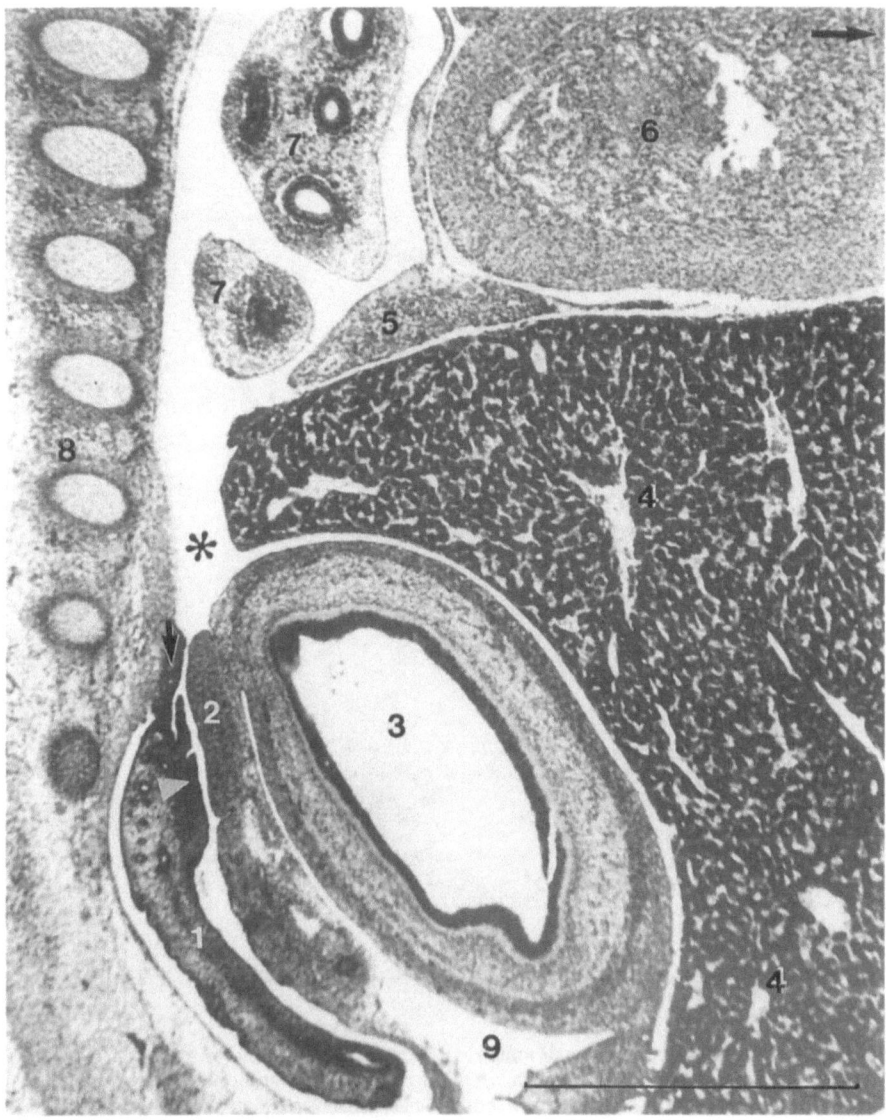

Fig. 2. Sagittal section through the right pleuroperitoneal canal of a 20-mm CRL female embryo. *1*, right mesonephros; *2*, spleen anlage; *3*, stomach; *4*, liver; *5*, anlage of diaphragm; *6*, heart; *7*, lung; *8*, dorsal body wall; *9*, bursa omentalis, the *asterisk* indicates the pleuroperitoneal canal, the *arrow* shows the cranial mesonephric ligament, the *arrowhead* indicates the gonad. *Bar*, 1 mm

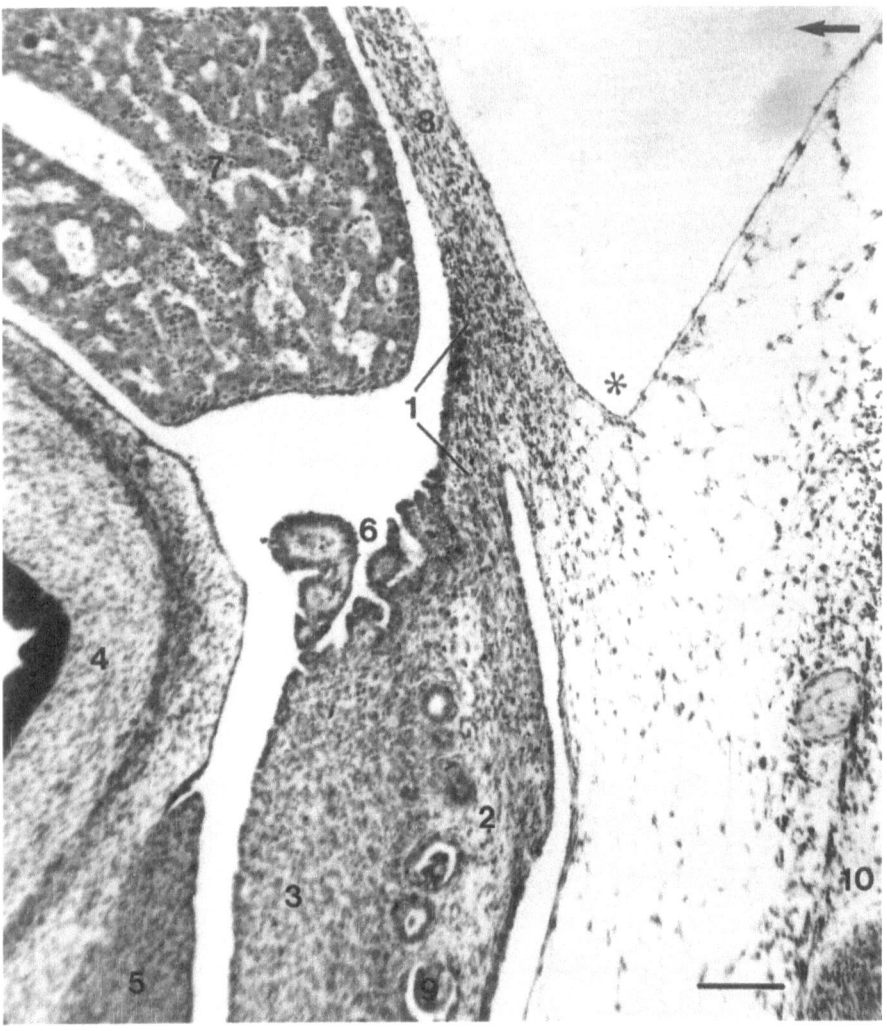

Fig. 3. Sagittal section through a female embryo of 21 mm CRL. *1*, cranial mesonephric ligament; *2*, cranial mesonephros; *3*, gonad; *4*, stomach; *5*, spleen; *6*, funnel field at the cranial part of müllerian duct; *7*, liver; *8*, anlage of diaphragm; *9*, mesonephric glomerulus; *10*, dorsal body wall. The *asterisk* shows the recessus costodiaphragmaticus. *Bar*, 0.1 mm

The dimension of the ligament may be well documented by scanning electron microscopy. Figure 6 shows its insertion into the border of the testis and the regressing mesonephros. Here, hydatides may develop as remnants of the müllerian or wolffian duct and the remodelling mesonephros. In the 70-mm male embryo (Fig. 7), a very thin ligament extends from the upper pole of the mesonephros to the dorsal part of the diaphragm.

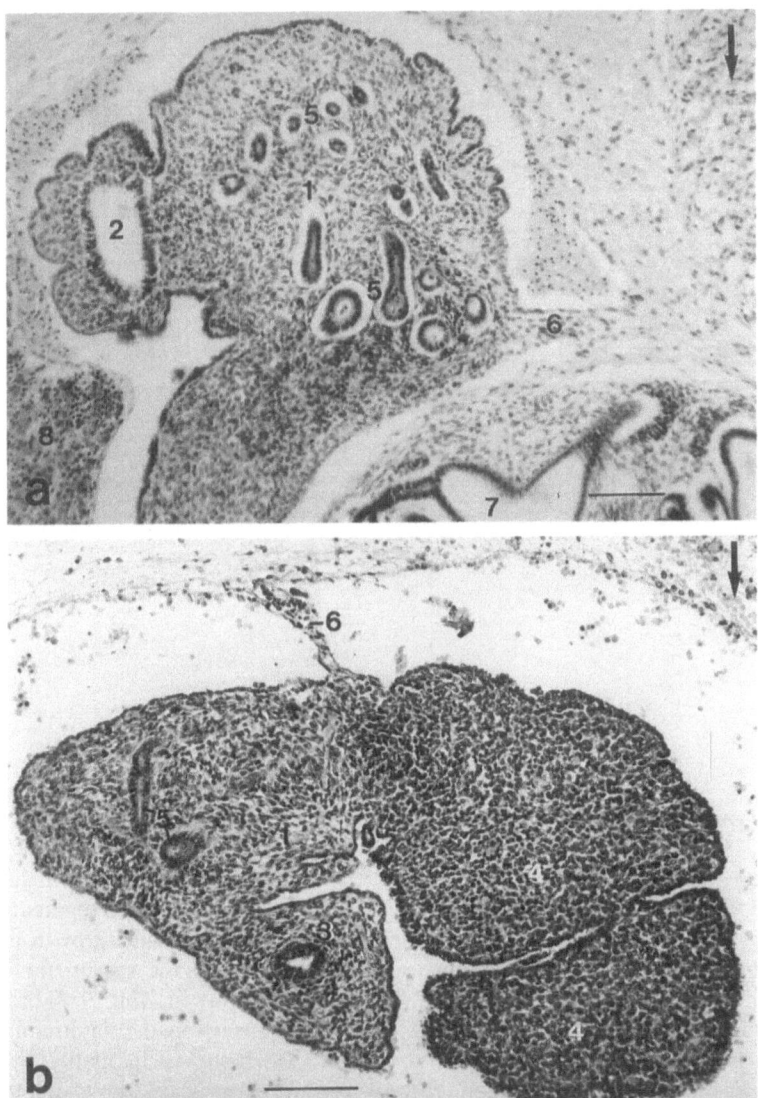

Fig. 4a-b. a Left mesonephros and ovary of a 29-mm CRL embryo. Frontal section. Note the wide abdominal ostium of the müllerian duct. **b** Female foetus 50 mm CRL. Horizontal section. Note the thin cranial mesonephric ligament. *1*, mesonephros; *2*, abdominal ostium of müllerian duct with anlage of fimbria; *3*, müllerian duct; *4*, ovary; *5*, tubuli of mesonephros; *6*, cranial mesonephric ligament; *7*, metanephros; *8*, liver. *Bars*, 0.1 mm

Thickness and length of the cranial mesonephric ligament measured by serial sections are summarised in Table 1. The average length of the ligament in these 13 embryos is 0.442 mm, the average thickness is 0.0356 mm. It is remarkable that the measurements do not reveal a notable increase during growth of the embryo.

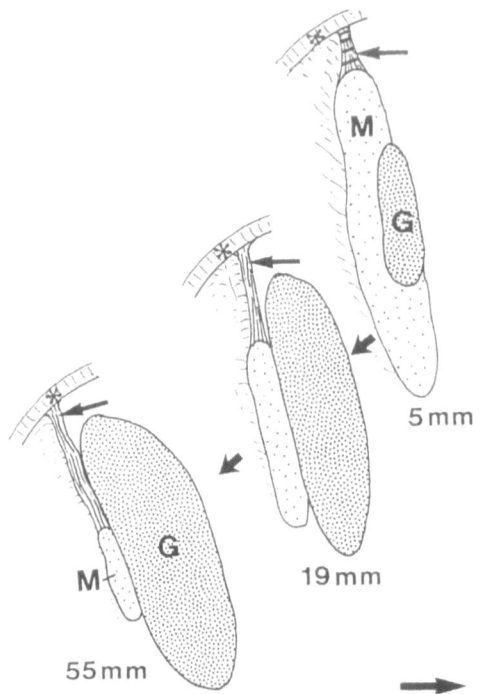

Fig. 5. Diagram showing the development of the cranial mesonephric ligament during regression of the mesonephros in 5–55-mm CRL embryos. Right lateral view. The mesonephros of the 5-mm embryo lies cranial in the pleuroperitoneal canal. The prospective ligament is indicated by *stripes. Asterisks* denote the anlage of diaphragm; *M,* mesonephros; *G,* gonad; *long arrows* show the cranial mesonephric ligament, *short arrows* show successive developing steps

Using our own data and those from the literature (Mall 1910; Felix 1911; Blechschmidt 1973; Gasser 1975), we checked the different positions of the diaphragm, mesonephros and gonad related to vertebral segments in embryos from 2.5 to 24 mm CRL (Table 2). Comparing the growth of the different notochord segments and assuming a constant length of thoracal segments no explosion-like growth of certain vertebral segments can be demonstrated which may be the reason for a relative caudal shifting of the gonads (Fig. 8). The schematic drawing (Fig. 9) shows the relationship between vertebral segments and dorsal insertion of diaphragm, mesonephros and gonad in embryos from 5 to 19 mm CRL. It shows clearly that the gonads in the early phase of development descent from level cervical 7/thoracal 8 to level thoracal 9/lumbar 3 (cranial pole prior to caudal) while the mesonephros is regressing. The gonads are indirectly attached to the anlage of the diaphragm (septum transversum) by the upper mesonephric ligament. Thus, we may conclude that with the descent of the diaphragm an inner descent of gonads indeed occurs during early development.

The level of origin of the testicular or ovarian artery is demonstrated in Fig. 10. From the third group of mesonephric arteries lying between meso- and metanephros, the last or second last will develop into the gonadal artery.

An inner descent of gonads already occurs during their development with coincident regression of the mesonephros. The ovaries reach their definite position, at the second or third sacral segments, corresponding to the level of the linea terminalis pelvis, very early in development. According to Felix (1911), this definite location may

10

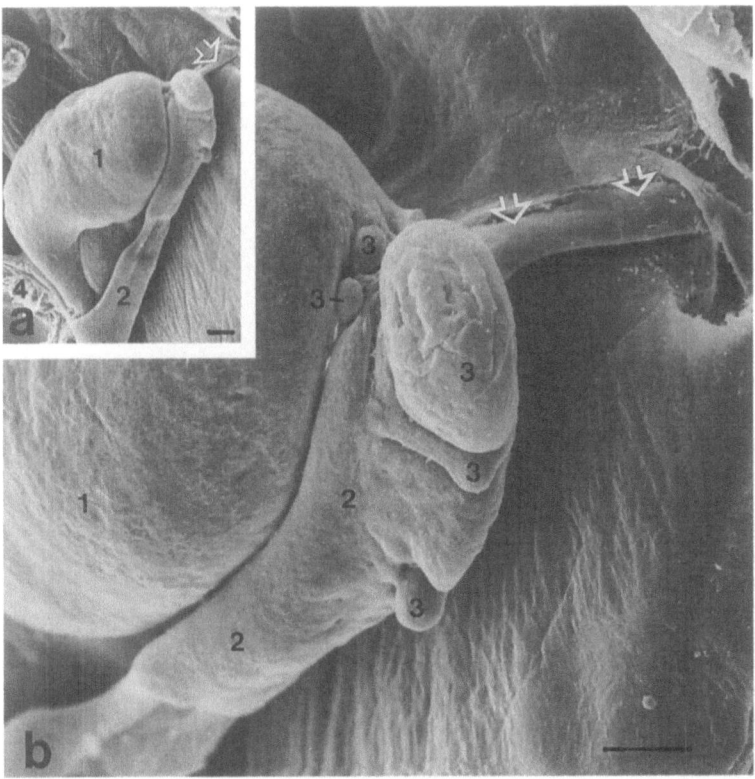

Fig. 6a,b. SEM micrographs of a 55-mm CRL male foetus. **a** Left gonad with both ducts, müllerian and wolffian. **b** Higher magnification of **a**, cranial view with cranial mesonephric ligament. *1*, testis; *2*, müllerian and wolffian ducts; *3*, anlage of hydatides; *4*, urinary bladder. The *open arrow* shows the cranial mesonephric ligament. *Bars*, 0.1 mm

be achieved in female embryos of 50 mm CRL. We found the gonads as early as in 25-mm embryos at a sacral position. The testes migrate further, and lie at the deep inguinal ring at 20 weeks (about 170 mm).

The inner descent of both gonads, male and female, depends on the descending septum transversum or anlage of the diaphragm. The latter is indirectly connected with the gonads by ligaments inserting into the border between the mesonephros and the gonad near the hydatid region. Therefore, these ligaments should be called cranial mesonephric ligaments (Blechschmidt 1960). The inconsistency of nomenclature of this paired ligament [(e.g. it is named diaphragmatic ligament in Clara (1967) and Patten (1968), suspensory ligament in Hamilton et al. (1972) or cranial gonadal ligament (Kollmann 1907; Drews 1993)] is discussed in detail by van der Schoot (1993c). According to Weber (1898) as cited by van der Schoot (1993c), the cranial gonadal ligament fuses with the remnants of the earlier mesonephric ligament after mesonephric regression. We did not find a ligament running from the cranial pole of

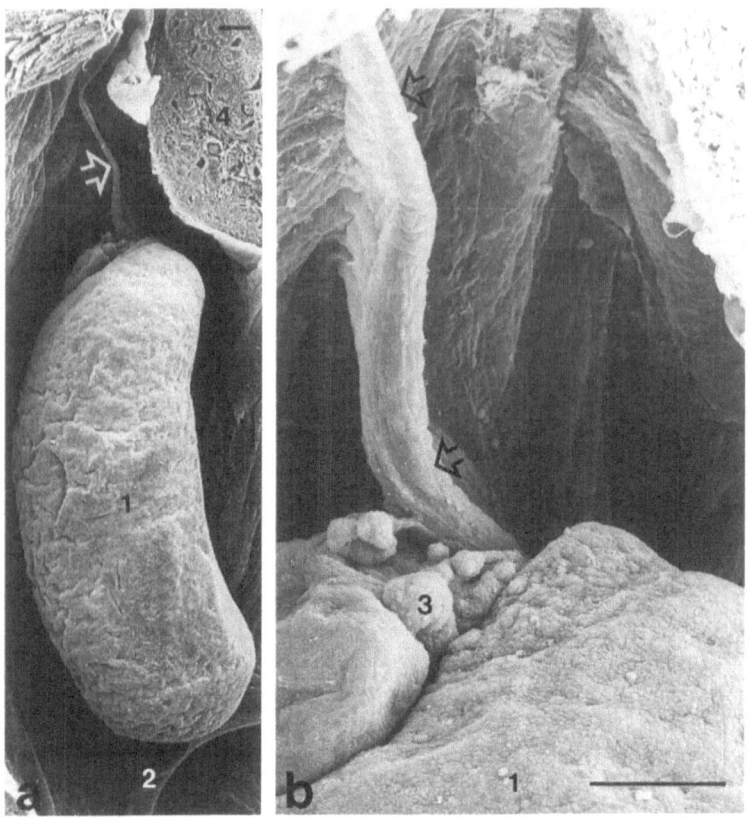

Fig. 7a,b. SEM micrographs of a 70-mm CRL male foetus. **a** Overall view of the right gonad with the cranial mesonephric ligament and the gubernaculum. **b** Higher magnification of **a**, with gonad and cranial mesonephric ligament. *1*, testis; *2*, gubernaculum; *3*, region of hydatides; *4*, metanephros. *Open arrows* show the cranial mesonephric ligament. *Bars*, 0.1 mm

the gonad to the developing diaphragm. But with the growth of the gonad relative to the mesonephros, the ligament of the latter comes into contact with the gonad (Fig. 5).

The cranial mesonephric ligament becomes thin and consists of loose mesenchyme (Felix 1911). Since it does not contain blood vessels, we agree with van der Schoot (1993c) that this ligament differs from the ovarian suspensory ligament of the adult, although many embryology textbooks state that it derives from the cranial mesonephric or gonadal ligament (e.g. Wartenberg 1994). The cranial mesonephric ligament is nothing other than a rudimentary structure (Horstmann and Stegner 1966; van der Schoot 1993c).

During further development, the position of gonads differs in both sexes. Van der Schoot (1993b) argues that the ovaries migrate craniolaterally with the ascending kidney due to the persistence of the cranial ovarian ligament. Androgens have been postulated to induce regression of the cranial "suspensory" ligament (van der Schoot and Elger 1992; Kersten et al. 1996; Emmen et al. 1998). Thus, exogenous androgens in

Table 1. Thickness and length of cranial mesonephric ligaments in 13 embryos

CRL (mm)	Gender	Length of ligament (mm)	Thickness of ligament (mm)
6.5	M	0.2 (insertion)	0.02 (insertion)
16.0	M	0.1 (insertion)	0.05 (insertion)
20.0	F	0.2 (insertion)	0.07
21.0	F	0.1	0.07
22.0	M	0.65	0.02–0.03
23.0	F	0.4–0.6	0.04
26.0	M	0.63	0.02
29.0	F	0.99	0.04
32.0	M	0.49	0.02
50.0	F	Torso	0.017
55.0	M	0.31	0.05
70.0	M	0.63	0.07
78.0	F	0.4–0.5	0.016

Table 2. Relations* of the diaphragm, mesonephros and gonads to vertebral segments in embryos of stages 9–21 CC

Embryo CRL (mm)	Diaphragm dorsal side	Mesonephros from cranial to caudal end	Gonads from cranial to caudal end
2.50	C2	Th2–Th5	–
4.25	C2	C6–L2	–
5.00	C4	C6–L2	C7–Th8
6.00	C5	*	*
7.00	C7	*	*
10.00	Th3	Th6–L1	Th4–Th12
10.50	Th1	Th2–L1	Th4–Th12
15.00	Th11	*	*
17.00	Th7	*	*
18.00	Th6	Th10–L3	Th10–L2
19.00	Th9	Th11–L3	Th9–L3
24.00	L1	*	*

*After Mall 1910; Felix 1911; Blechschmidt 1973; Gasser 1975.

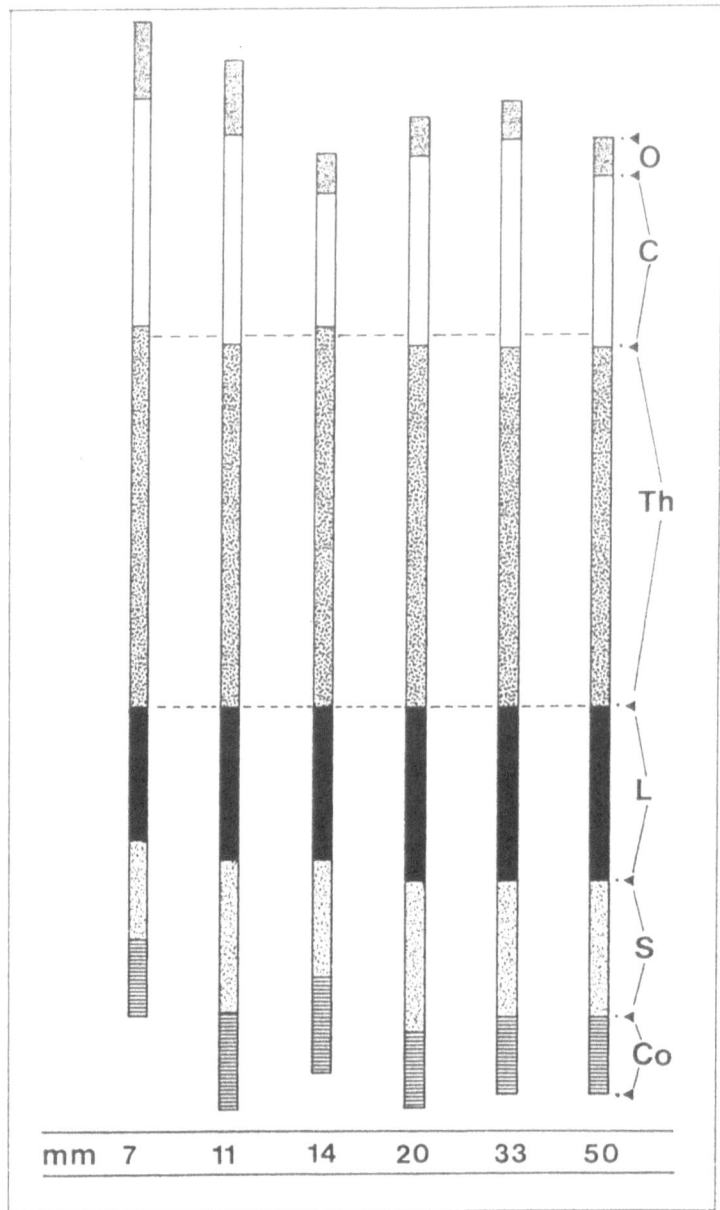

Fig. 8. Comparison of proportions of notochordal segments in embryos of 7–50 mm CRL assuming an equal length of the thoracal segment. *O*, occipital; *C*, cervical; *Th*, thoracal; *L*, lumbar; *S*, sacral; *Co*, coccygeal. (After Bardeen 1910)

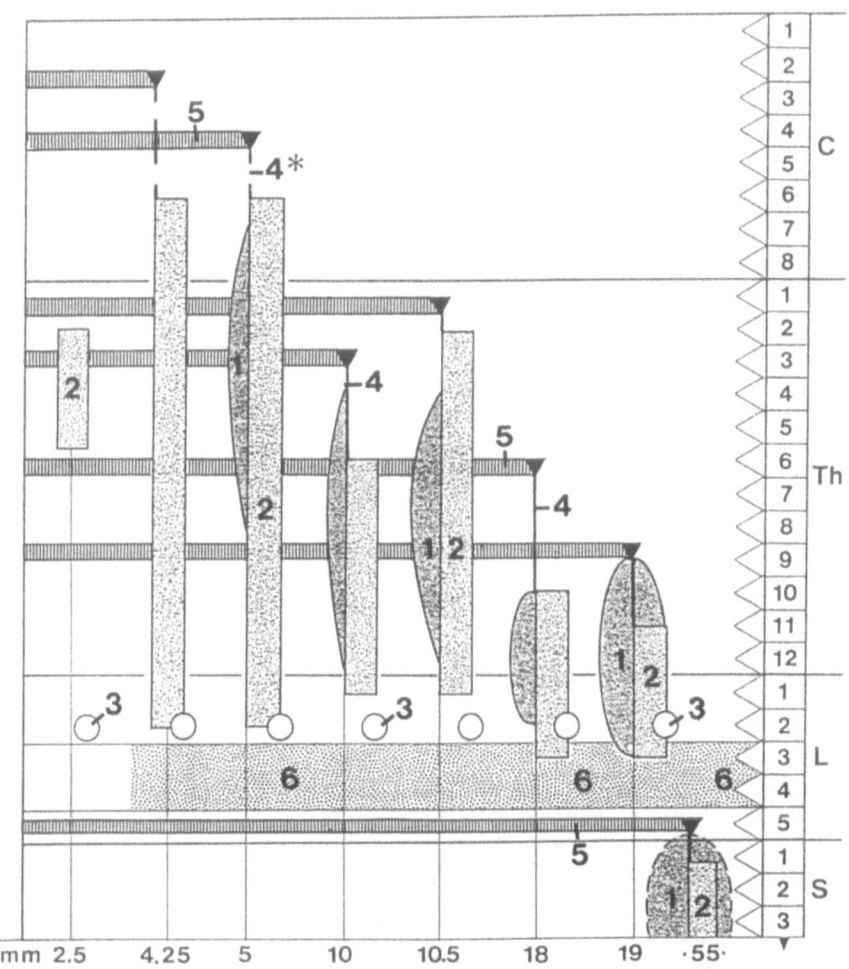

Fig. 9. Changing position of the dorsal insertion of the diaphragm anlage, mesonephros and gonad to vertebral segments in human embryos of stages 9–21 Carnegie collection (CC) (2.5–19 mm CRL) and a foetus of 55 mm CRL. Note the level of origin of ovarian or testicular arteries as well as the inferior mesenterial artery. *1*, gonad; *2*, mesonephros; *3*, level of origin of testicular or ovarian artery; *4*, cranial mesonephric ligament; *4**, cranial mesonephric ligament not yet formed, cranial mesonephros lying in the pleuroperitoneal canal; *5*, anlage of diaphragm; *6*, origin of inferior mesenterial artery, *C,* cervical; *Th,* thoracal; *L,* lumbar; *S,* sacral. (After Mall 1910; Felix 1911; Blechschmidt 1973; Gasser 1975)

female mice cause the ovary to be more mobile and lower descent into the abdominal cavity (Lee and Hutson 1999). However, according to Neumann (1933), the descent of ovaries to the inguinal region is hindered because they lie behind the müllerian duct and they are fixed by the ligamentum ovarii proprium.

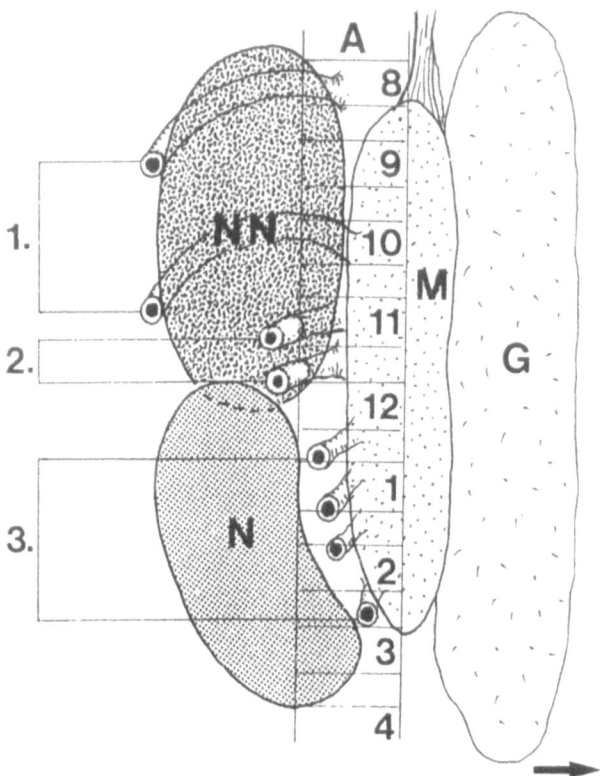

Fig. 10. Origin of the testicular/ovarian artery and the renal artery. The *marks* on the aorta indicate the level of vertebral bodies and intervertebral discs thoracal 8–12 and lumbar 1–4. Right lateral view. *1,* group of mesonephric arteries running dorsal to suprarenal gland; *2,* group of mesonephric arteries running through suprarenal gland; *3,* group of mesonephric arteries between meso- and metanephros; *A,* aorta; *G,* gonad; *M,* mesonephros; *N,* metanephros; *NN,* suprarenal gland. The *arrow* indicates ventral direction. (Modified from Felix 1911, who projected the reconstruction of an 18-mm CRL embryo into a 19.4-mm CRL embryo)

The biphasic model for testicular descent postulated by Hutson (1985) predicts that the first transabdominal phase of testicular descent is regulated by müllerian inhibiting substance (MIS), but studies with animals immunised against MIS argue controversially (Tran et al. 1986).

According to our results, the ovarian suspensory ligament is a de novo formation which develops with the vascularisation of the ovaries. It does not derive from the cranial gonadal ligament (which has been named cranial mesonephric ligament). This ligament is a transitory structure in both sexes. Therefore, the discussion about growth involution of this ligament by anti-müllerian hormone is obsolete.

Another role of MIS in testicular descent was proposed by Siow and Fallat (1997), i.e. interacting with epidermal growth factor. Finally, Clarnette et al. (1997) suggested that MIS controls the swelling reaction in the male gubernaculum, which seems to be responsible for the first phase of testicular descent (see for discussion Hutson and Beasley 1992; Hutson et al. 1990, 1996, 1997). This will be discussed in Sect. 4.2.1.

4 Development, Shape and Fate of Gubernaculum Hunteri and Processus Vaginalis Peritonei – Own Phases of Testicular Descent

4.1 Results

Testicular descent is divided into an inner and outer one (Hutson 1985) or into three stages: abdominal, canalicular, and scrotal descent (e.g. Heyns 1987; Sampaio and Favorito 1998). These classifications do not seem to be sufficiently detailed with respect to clinical (e.g. cryptorchidism) or experimental work (e.g. function of MIS or androgens). Therefore, we have divided testicular descent into five phases referring to significant processes mainly in the development of the gubernaculum and position of testes.

4.1.1 Phase I: Early Development of the Gubernaculum

In stage 13–14 Carnegie collection (CC) embryos (5–7 mm CRL, 5 weeks), shortly after the wolffian ducts have reached the cloaca and the ureteric buds have appeared, the caudal part of the mesonephric fold contacts a mesenchymal elevation, the conus inguinalis, which bulges out of the abdominal wall into the abdominal cavity (Fig. 11). This occurs at the transition from a longitudinal to a transverse position of the mesonephric fold, that means the first buckling of this fold (Felix 1911) and later the crossing point of müllerian and wolffian duct.

With the punctuated contact between mesonephros and abdominal wall (Fig. 12, stage 18 CC), the abdominal gubernaculum is established and the position of the inner inguinal ring is defined.

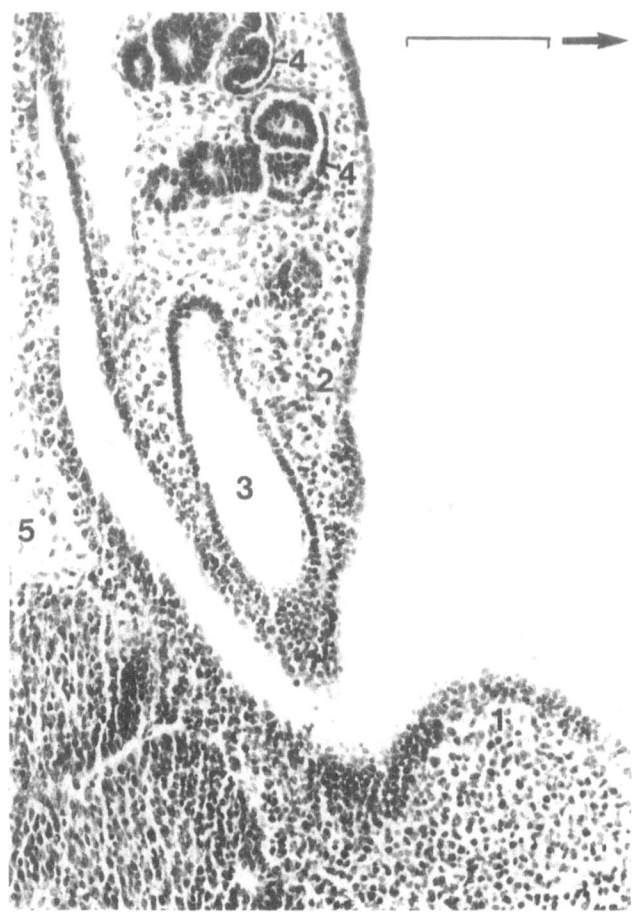

Fig. 11. Sagittal section through the caudal part of the plica mesonephrica, conus inguinalis and ventral abdominal wall of a 6.5-mm CRL embryo (stage 14 CC). *1,* conus inguinalis; *2,* plica mesonephrica; *3,* wolffian duct; *4,* glomeruli of mesonephros; *5,* dorsal body wall. *Bar,* 0.1 mm

4.1.2
Phase II:
Different Parts of the Gubernaculum, Appearance of Processus Vaginalis

At stage 20 CC (7 weeks, 21 mm), the gubernaculum has formed a broad plate. The hinge-like adhesion point is engraved in the mesenchymal surrounding of both ducts – müllerian and wolffian – at the area where their crossing over occurs (Figs. 13, 14). The topographical relationships between the abdominal gubernaculum, genital ducts, mesonephros and gonad are shown in Fig. 15, demonstrating that the gubernaculum is not in contact with gonad at early stages.

During stages 20–23 CC, the abdominal gubernaculum is broad in its proximal section and forms a round cord further distally (Fig. 16a). An interstitial part is

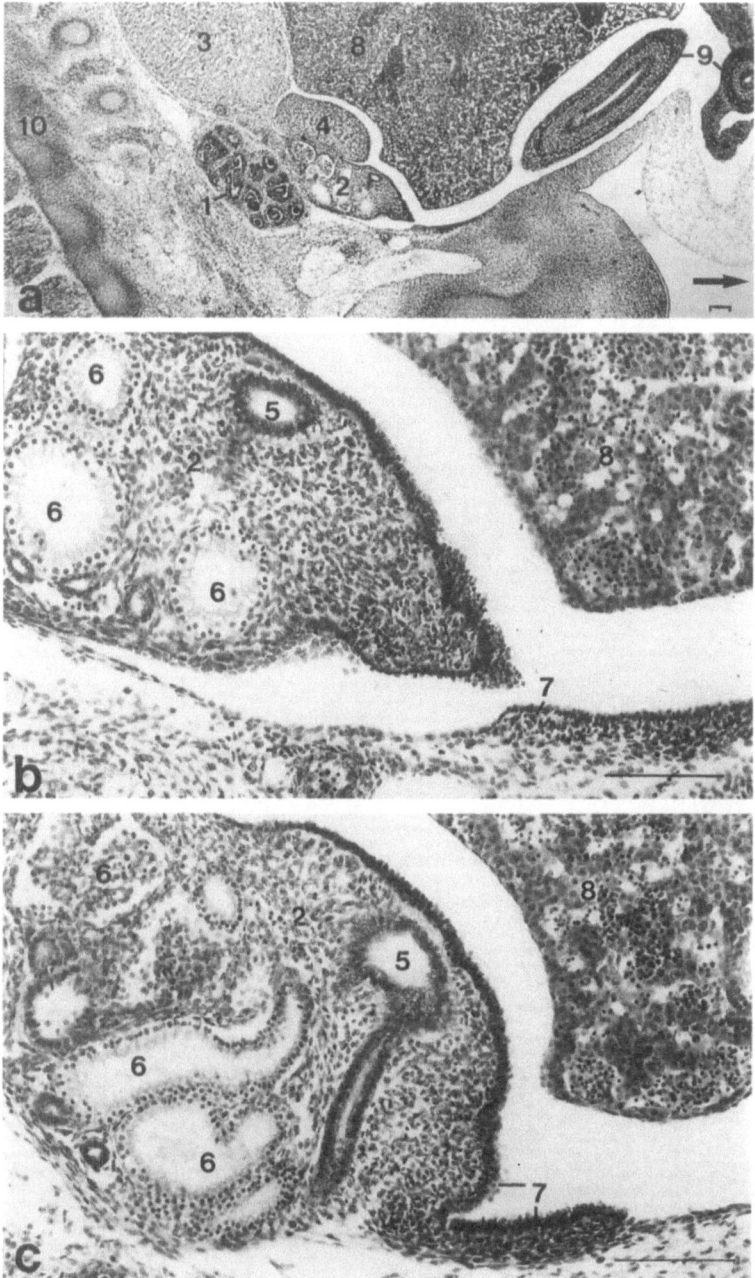

Fig. 12a–c. Sagittal sections through a 16-mm CRL embryo (stage 18 CC) after the contact between mesonephros and ventral abdominal wall has been established. **a** Overall view. **b** Detail of **a. c** Lateral section of **a.** *1*, metanephros; *2*, mesonephros; *3*, suprarenal gland; *4*, testis; *5*, mesonephric duct; *6*, mesonephric tubuli and glomeruli; *7*, abdominal gubernaculum/mesenchymal condensation at ventral body wall and mesonephros; *8*, liver; *9*, intestine; *10*, vertebral anlage. *Bars*, 0.1 mm

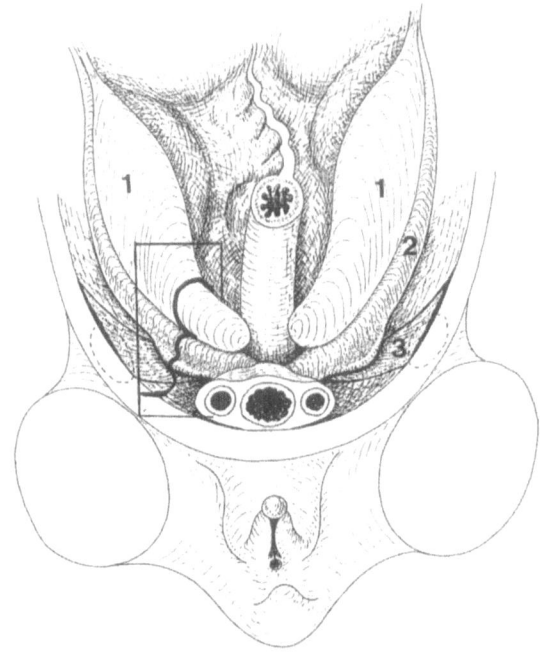

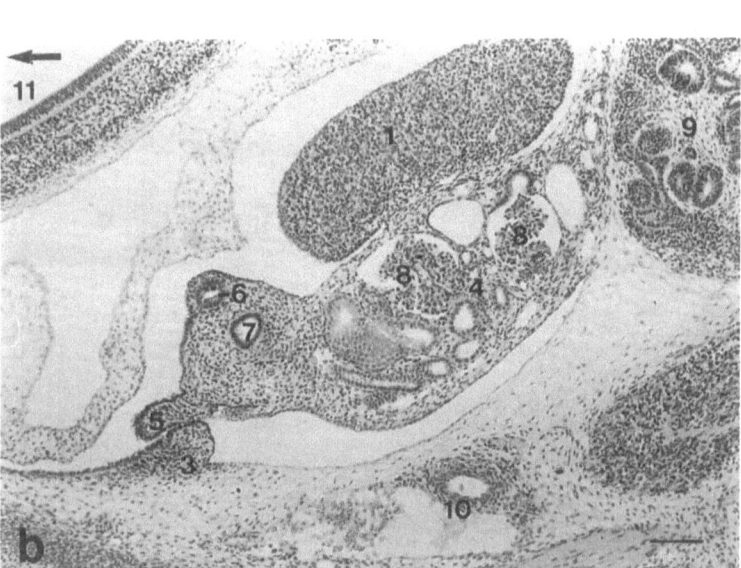

Fig. 13. a Genital region and caudal part of the opened abdomen. 21 mm CRL female embryo, stage 20 CC. The plane of section in **b** is indicated by a *rectangle*. **b** Sagittal section through the right mesonephros. Note the hinge-like link between the abdominal gubernaculum and free end (nose) of the caudal mesonephros. *1*, ovary; *2*, mesenchyme of the genital ducts; *3*, abdominal gubernaculum; *4*, mesonephros; *5*, free end of caudal mesonephros; *6*, müllerian duct; *7*, wolffian duct; *8*, glomeruli of mesonephros; *9*, metanephros; *10*, right inferior epigastric vessels; *11*, stomach. *Bar*, 0.1 mm

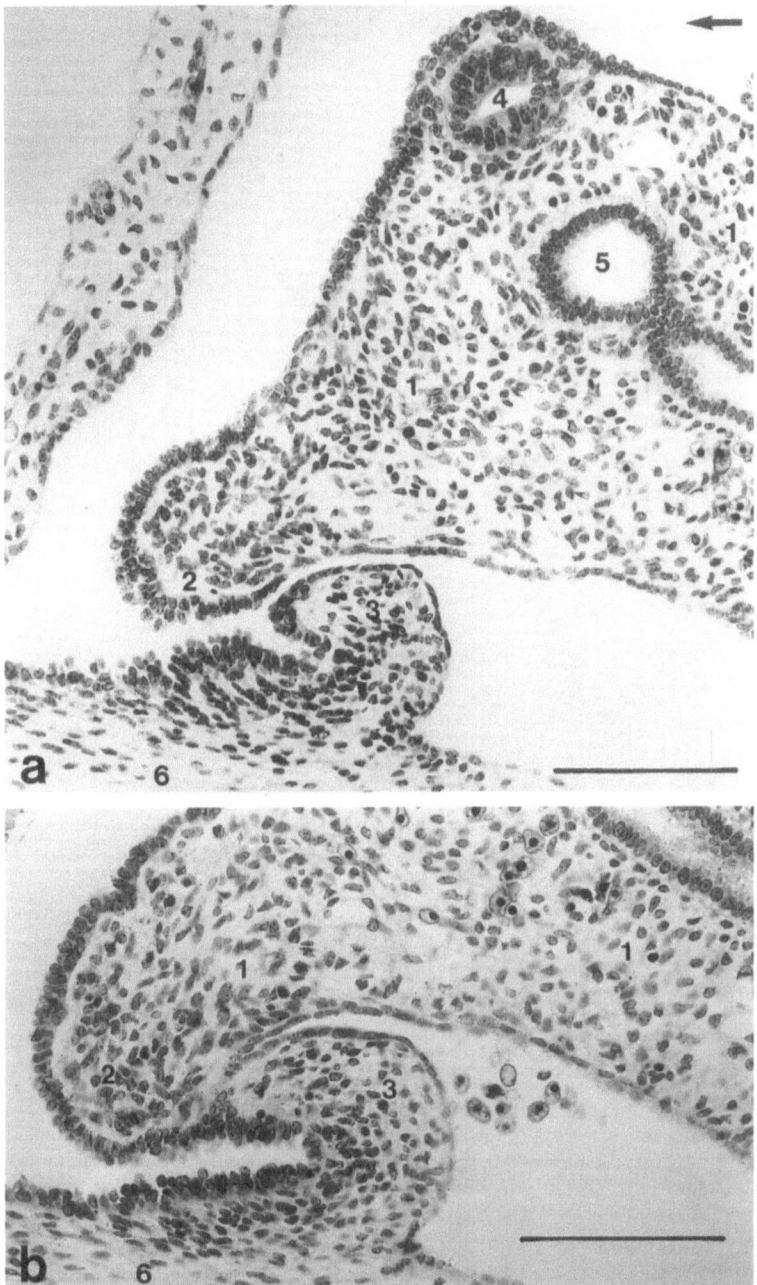

Fig. 14a,b. Detail of contact between ventral abdominal wall and mesonephros. Same embryo as in Fig. 13. **a** Conus inguinalis near the "nose" of mesonephros. **b** Hinge-like link between the conus inguinalis and the caudal part of mesonephros. The abdominal gubernaculum is now established. *1*, mesonephros; *2*, caudal part of mesonephros; *3*, abdominal gubernaculum/Conus inguinalis; *4*, müllerian duct; *5*, wolffian duct; *6*, ventral abdominal wall. *Bars*, 0.1 mm

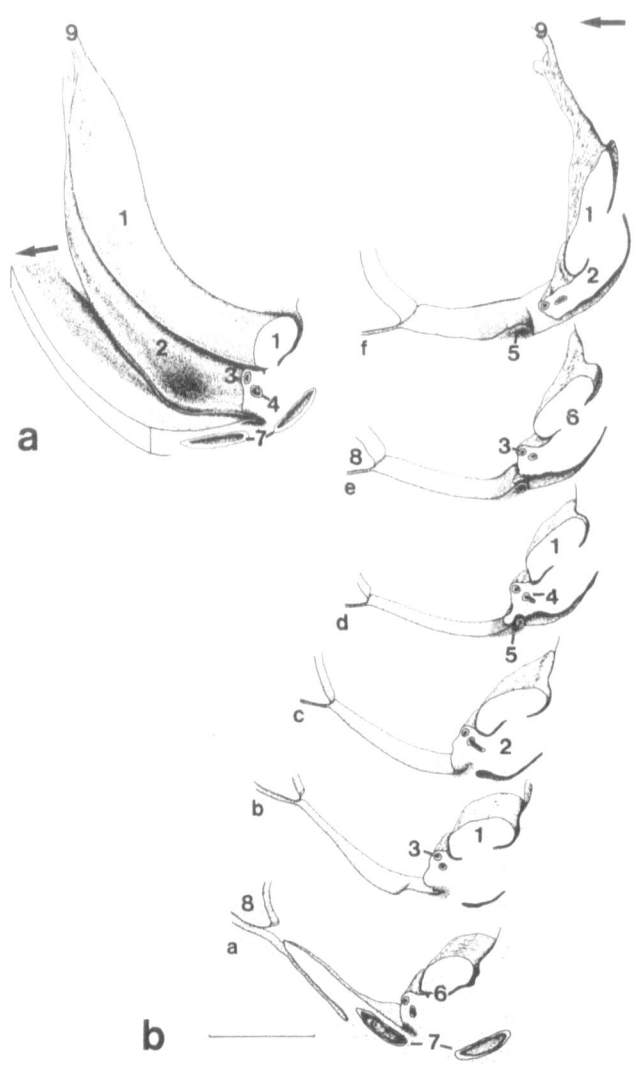

Fig. 15a-b. a Reconstruction of the right gonad and mesonephros of the 21-mm CRL female embryo of Figs. 13 and 14 to show the formation of the abdominal gubernaculum. **b** Sectioned from medial *(a)* to lateral *(f)* side. *1*, ovary; *2*, mesonephros; *3*, müllerian duct; *4*, wolffian duct; *5*, abdominal gubernaculum; *6*, mesovarium; *7*, right umbilical artery; *8*, liver; *9*, cranial mesonephric ligament. *Bar*, 1 mm

formed penetrating the abdominal muscle layers. The gubernaculum ends with a filamentous subcutaneous part in the pubic region.

Figure 16 a–c shows the abdominal part of the gubernaculum from a 22-mm CRL male embryo (stage 21 CC) which is surrounded by peritoneum except where its

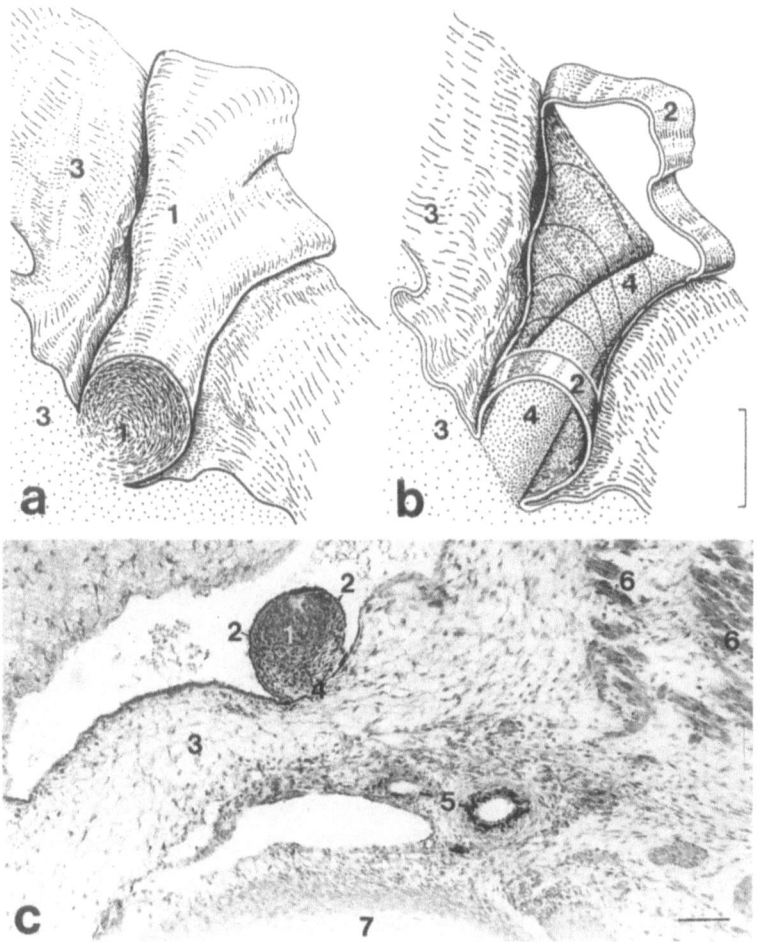

Fig. 16a–c. Right abdominal gubernaculum of a 22-mm CRL male embryo. **a** Reconstruction with parietal peritoneum. **b** Reconstruction showing the isolated peritoneal coat of the gubernaculum and its adhesion zone with ventral body wall. *Bar,* 0.1 mm. **c** Histological section stained with anti-smooth muscle actin. *Bar,* 0.05 mm. *1,* abdominal gubernaculum; *2,* peritoneal coat; *3,* ventral abdominal wall; *4,* adhesion zone of abdominal wall and gubernaculum; *5,* inferior epigastric vessels with stained muscle layer; *6,* abdominal muscles; *7,* cartilaginous anlage of pubic bone

posterior surface is attached at the abdominal wall. Though the gubernaculum widens proximally, the adhesion zone is as narrow as it is distally. Thus, the dorsal side of the proximal part is also partly covered by peritoneum (Fig. 16b). The abdominal gubernaculum is rich in cells (Fig. 16c) which are highly proliferating as seen by the proliferation marker PCNA (not shown) but does not yet contain smooth muscle cells.

During this period also, the processus vaginalis appears – a blind pouch extending from the abdominal cavity into the inguinal region lying close to the interstitial gubernaculum (Fig. 16a–c). The processus vaginalis is lined by peritoneal epithelium

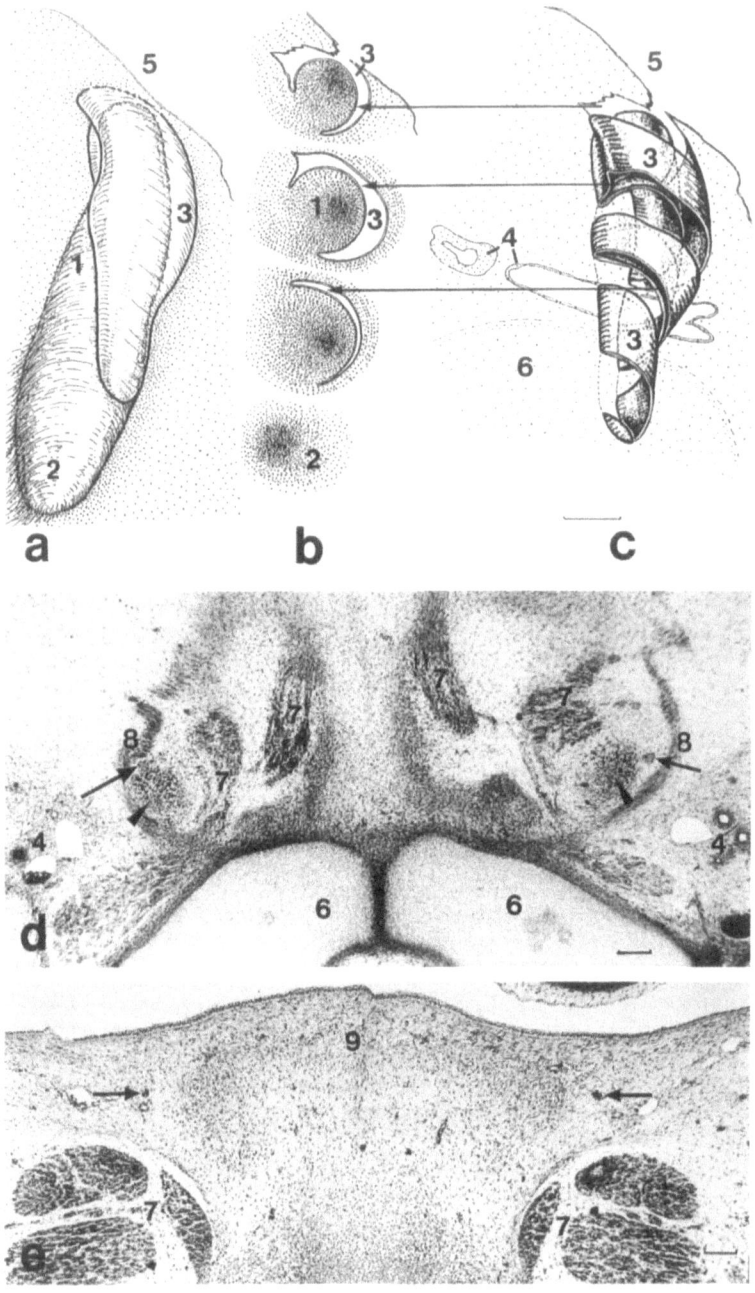

Fig. 17a–e. Reconstruction of the right interstitial and subcutaneous gubernaculum with formation of the processus vaginalis peritonei. Same embryo as in Fig. 16, frontal sections. **a** Processus vaginalis adjacent to the interstitial gubernaculum. **b** Sections through the inguinal region. **c** Isolated coat of processus vaginalis peritonei. The levels of sections in **b** are indicated. **d** Frontal section through distal part of both interstitial gubernacula with related genital branch of the genitofemoral nerve and aponeurosis of external oblique muscle. **e** Section through the cranial part of the scrotum with the

part of the scrotum with the genital branch of the genitofemoral nerve. *1*, interstitial gubernaculum; *2*, subcutaneous gubernaculum; *3*, processus vaginalis peritonei; *4*, right inferior epigastric vessels; *5*, abdomen; *6*, cartilaginous anlage of symphysis pubis; *7*, abdominal musculature (**d**)/thigh musculature (**e**); *8*, external oblique muscle with aponeurosis; *9*, cranial part of scrotum. The *arrowhead* indicates the interstitial gubernaculum, the *arrow* shows the genital branch of the genitofemoral nerve. *Bars*, 0.1 mm

which is dorsally firmly attached to the ventral side of the interstitial part of gubernaculum. Most of the interstitial gubernaculum, therefore, has a retroperitoneal position.

The genital branch of the genitofemoral nerve accompanies and precedes the gubernaculum (Fig. 17d,e).

A high cell density is also characteristic for the interstitial part of the gubernaculum (Fig. 18a) whereas the subcutaneous part consists of looser mesenchyme and a coat of cremaster muscle fibres deriving from the inner layers of abdominal muscles (Fig. 18b).

The early stages of gubernaculum development are similar in both sexes: the cell-rich part of the gubernaculum attaches at the mesenchymal layer of the genital ducts (Fig. 19).

In the 26-mm CRL male embryo (stage 22 CC), the caudal pole of the testis is attached to the dorsal mesenchyme of the genital ducts by a peritoneal fold (Fig. 20) which is found medial and opposite to the abdominal gubernaculum. Thus, an indirect connection between the caudal pole of the testis and the ventral abdominal wall is established. In female embryos, the ligamentum ovarii proprium is formed in the same way, and the lateral angles of the uterine cavity are determined (Fig. 21; see also Figs. 50, 51). The partial reconstruction and the diagrams in Fig. 22 illustrate these relationships. The voluminous mesenchyme of the müllerian duct and the wolffian duct (duct of Gartner) distinctly separates the gubernaculum from the gonads in both sexes. The complete course, shape and relations to the processus vaginalis and the genitofemoral nerve in the 29-mm CRL female embryo are shown in Fig. 23a,b. The well-established genital branch of the genitofemoral nerve is connected within the dorsal part of the interstitial gubernaculum. It always runs parallel to the curved gubernaculum and the processus vaginalis peritonei attached at the ventromedial side of the latter.

4.1.3
Phase IIa:
Growth of Gubernaculum and Processus Vaginalis Peritonei

This phase includes embryos and foetuses from 32 to 55 mm CRL, 8–10 weeks. This phase is characterised by an enormous increase of volume and length of the gubernaculum as well as the processus vaginalis. In the 32-mm CRL male embryo (Fig. 24), the abdominal gubernaculum forms a round cord wrapped in parietal peritoneum. Craniomedially, it is anchored within the mesenchyme of the genital ducts. In the middle it forms a free ridge at the inner side of the abdominal wall. Near the internal

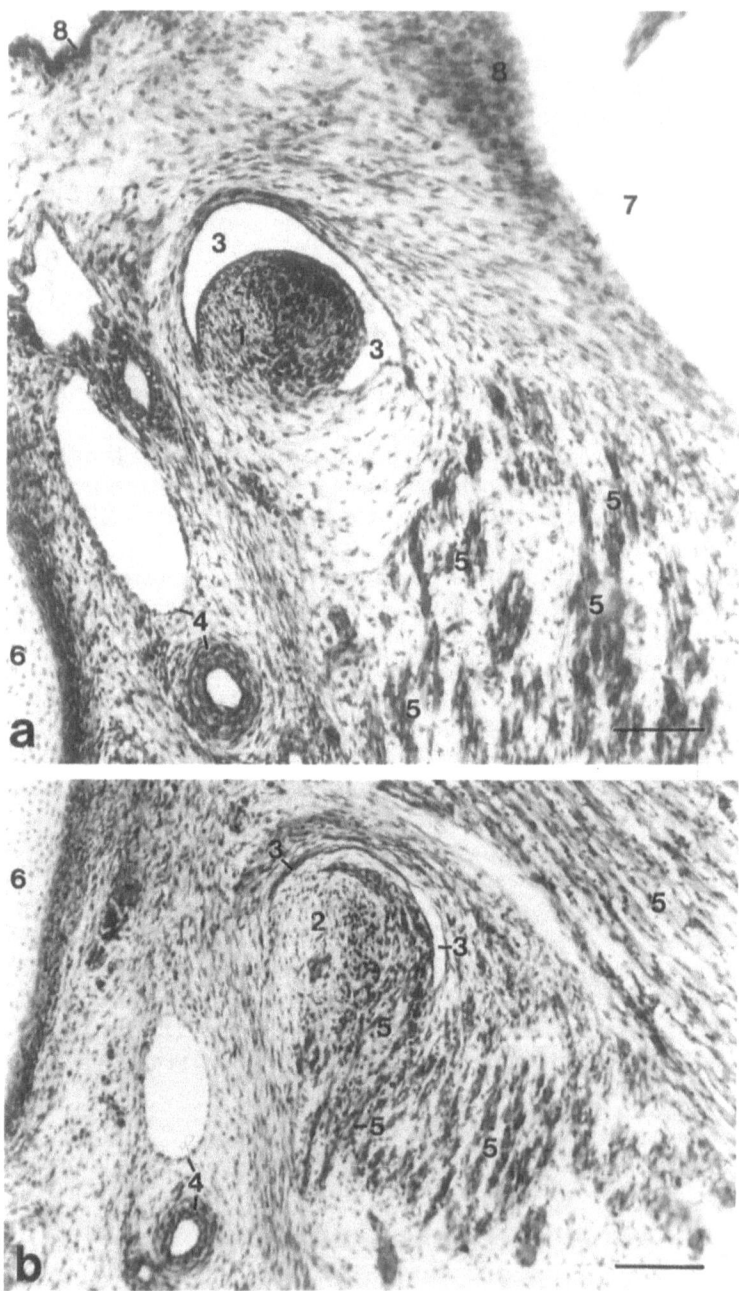

Fig. 18a,b. Frontal sections of the same embryo as in Figs. 16 and 17. **a** Right interstitial gubernaculum. **b** Subcutaneous gubernaculum. *1*, interstitial gubernaculum; *2*, interstitial gubernaculum more distally; *3*, processus vaginalis; *4*, right inferior epigastric vessels; *5*, abdominal musculature; *6*, cartilaginous anlage of pubic bone; *7*, abdominal cavity; *8*, peritoneum. *Bars*, 0.1 mm

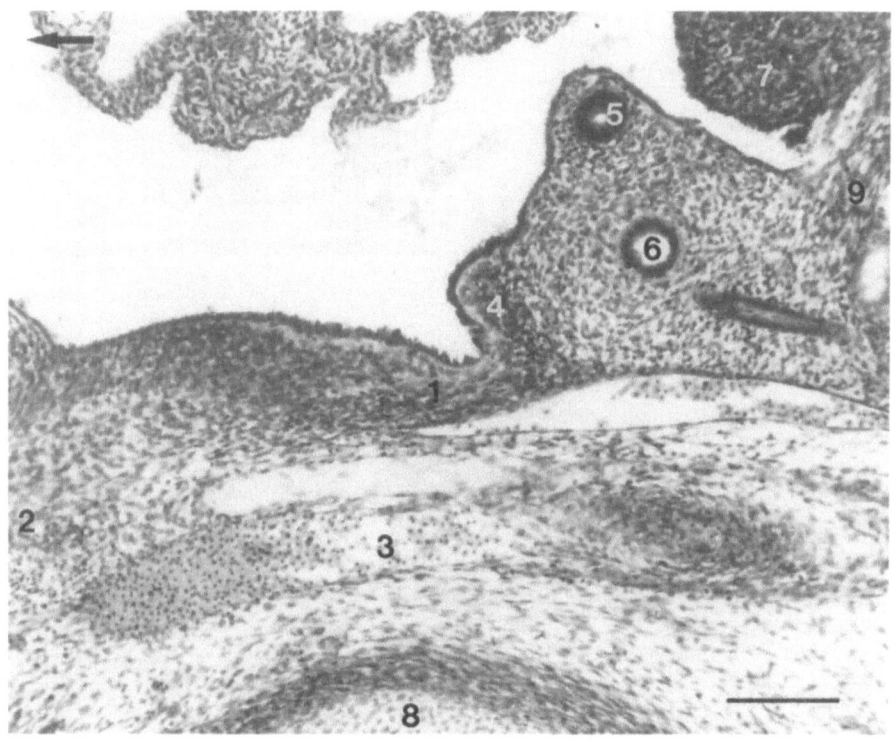

Fig. 19. Sagittal section of a 23-mm CRL female embryo. *1*, abdominal gubernaculum; *2*, interstitial gubernaculum; *3*, right umbilical artery; *4*, cells of gubernaculum at mesenchyme of genital ducts; *5*, müllerian duct; *6*, wolffian duct; *7*, ovary; *8*, pubic bone; *9*, mesovarium. *Bar*, 0.1 mm

inguinal ring, the gubernaculum broadens before penetrating the abdominal wall. The processus vaginalis is found ventrally. No direct contact to the testis exists.

The well-developed gubernaculum in a female foetus (anlage of round ligament of uterus) has a fibrous structure with a dense core. It is accompanied by the arteria ligamenti teretis uteri and the genitofemoral nerve (Fig. 25).

In a 45-mm CRL male embryo, the retroperitoneal abdominal gubernaculum forms a narrow ridge elevating immediately lateral to the plica of the inferior epigastric vessels (Fig. 26a,b). The parietal peritoneum exhibits a deeper excavation between the gubernaculum and the vessels than at the lateral side of the gubernaculum. Thus, the processus vaginalis becomes medially more distinct extending into the abdominal wall (Fig. 26c). The interstitial gubernaculum (Fig. 27) broadens into loose mesenchyme descending through the external oblique aponeurosis. The processus vaginalis forms a blind pouch at the ventromedial side of the proximal part of the interstitial gubernaculum. The cell-rich abdominal gubernaculum and the core of the interstitial gubernaculum are shown in Fig. 28. Caudal fibres of the inner abdominal muscles contact the interstitial gubernaculum and turn to the proximal part of the latter (Fig. 28b). The distal part penetrates the external abdominal muscle (Fig. 28a) and reaches as subcutaneous gubernaculum the symphyseal region.

27

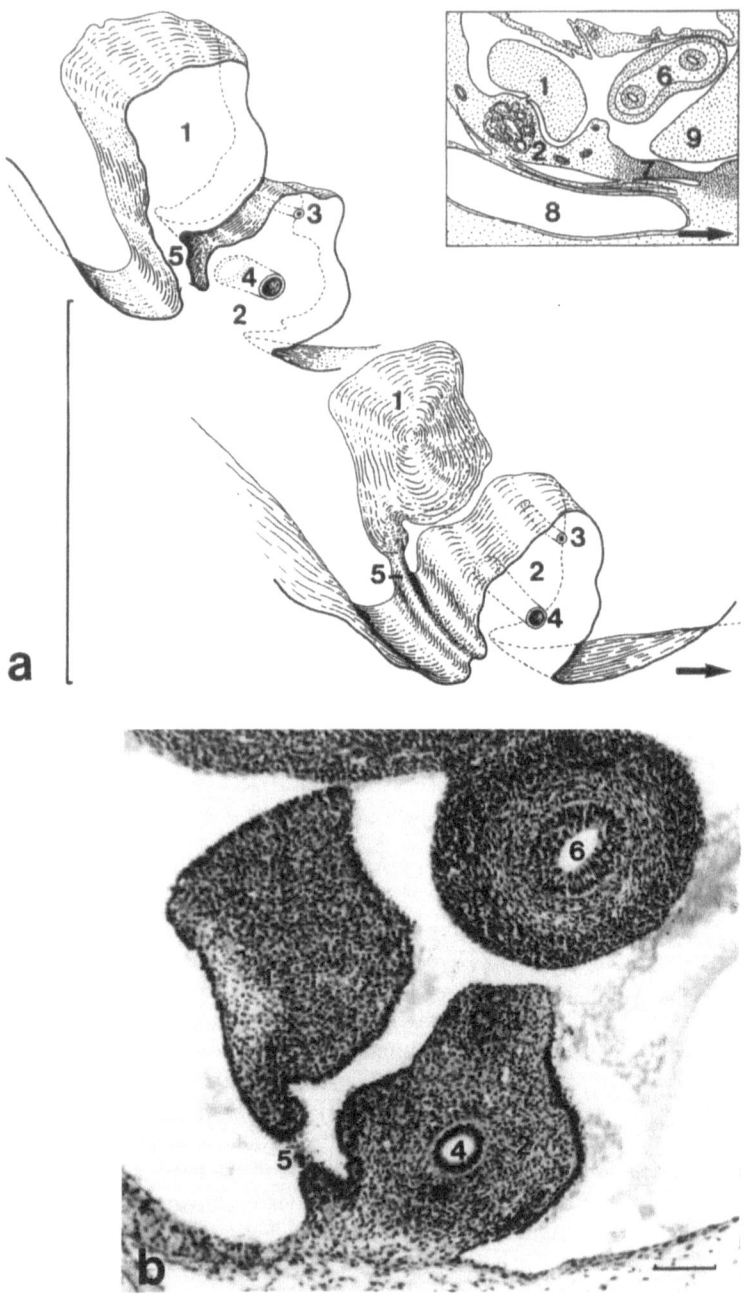

Fig. 20 a,b

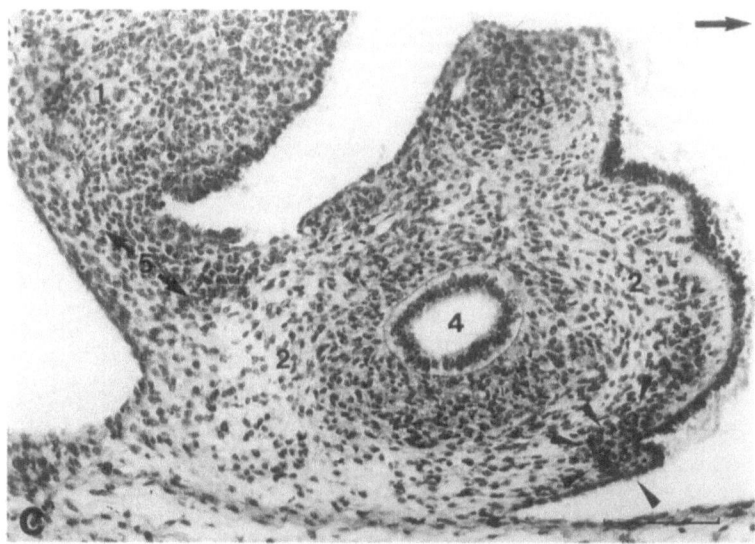

Fig. 20a–c. Connection between caudal pole of the testis and mesenchymal surrounding of genital ducts in a 26-mm CRL male embryo (stage 22 CC) **a** Bipartite reconstruction block. Medial view on caudal pole of left testis connecting the dorsal duct mesenchyme by a meso. Inset: abdominal gubernaculum. Distance from section b is 91 μm more lateral. **b** Sagittal section showing connection between the caudal pole of testis and dorsal mesenchyme of genital ducts. **c** Further lateral, contacts of abdominal gubernaculum with duct mesenchyme. *1,* caudal pole of testis; *2,* mesonephros; *3,* regressing müllerian duct; *4,* wolffian duct; *5,* connection of caudal pole of testis with mesenchyme bed of genital ducts; *6,* intestine; *7,* abdominal gubernaculum; *8,* left umbilical artery; *9,* liver. *Arrowheads* show the insertion of abdominal gubernaculum. *Bars* **a,** 1.0 mm; **b, c,** 0.1 mm

As demonstrated in Fig. 29a, also in female embryos of this age (about 50 mm CRL), the gubernaculum or anlage of the round ligament of the uterus is a mighty, cell-rich structure inserting at the mesenchymal layer of the müllerian duct. It is not continuous with the ligamentum ovarii proprium or mesovarium. The ovary is fixed by the mesovarium and the ovarian ligament with the back of the anlage of broad ligament. Therefore, no direct connection between gonad and gubernaculum exists.

Immunostaining with anti-smooth muscle actin shows some smooth muscle cells especially concentrated near the epigastric vessels beneath the processus vaginalis peritonei (Fig. 29b,c).

4.1.4
Phase III:
Growth of Testes and Change of Relations

The growth and relations of gonads can be best studied with scanning electron microscopy (SEM; Fig. 30). In the 55-mm CRL male foetus, the testes have increased in volume while the mesonephros and the müllerian duct are regressing. Since testicular

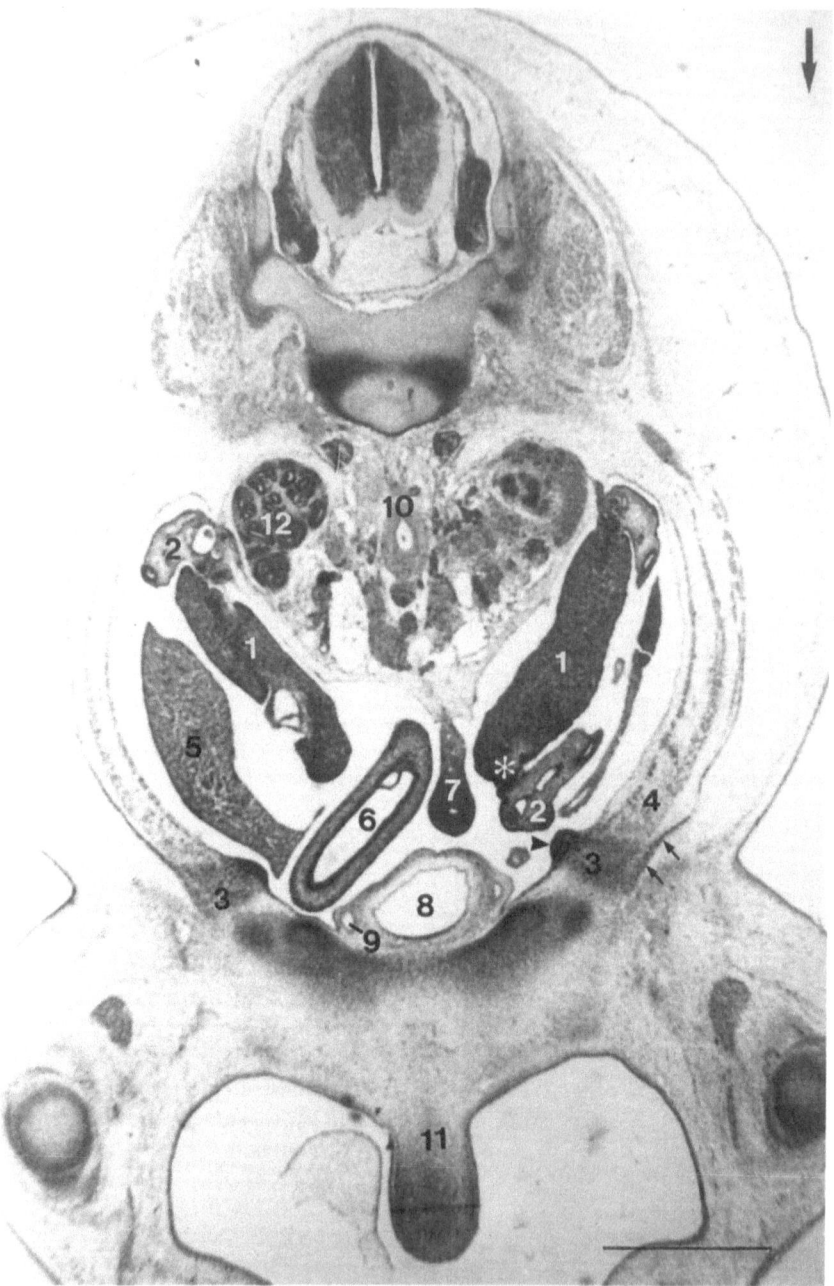

Fig. 21. Transverse section of a 26-mm CRL stage 22 CC female embryo. *1*, ovary; *2*, mesonephros; *3*, interstitial gubernaculum; *4*, abdominal musculature; *5*, liver; *6*, large intestine; *7*, rectum; *8*, urinary bladder; *9*, right umbilical artery; *10*, aorta; *11*, genital tubercle; *12*, metanephros. The *arrowhead* shows the abdominal gubernaculum, *arrows* show aponeurosis of external oblique muscle, the *asterisk* indicates the dorsal superior angle (Tubenwinkel) of the uterus with ligamentum ovarii proprium. *Bar,* 1 mm

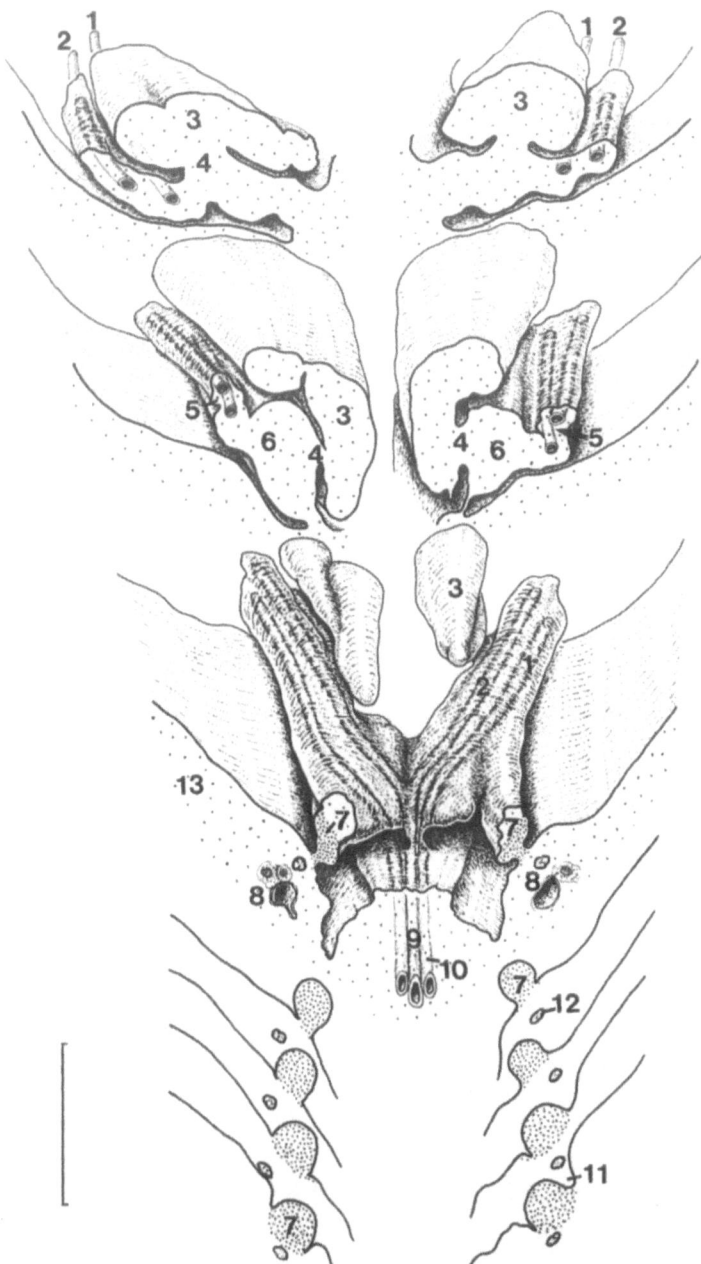

Fig. 22. Tripartite reconstruction and drawing of sections through the abdominal gubernaculum with ovaries and the crossing genital ducts of a 29-mm CRL female embryo, ventral view. *1,* wolffian duct; *2,* müllerian duct; *3,* ovary; *4,* mesovarium; *5,* crossing zone of müllerian and wolffian duct; *6,* mesonephros; *7,* abdominal gubernaculum; *8,* inferior epigastric vessels; *9,* anlage of uterus; *10,* Gartner's duct; *11,* developing processus vaginalis; *12,* genital branch of genitofemoral nerve; *13,* abdominal wall. *Bar,* 1 mm

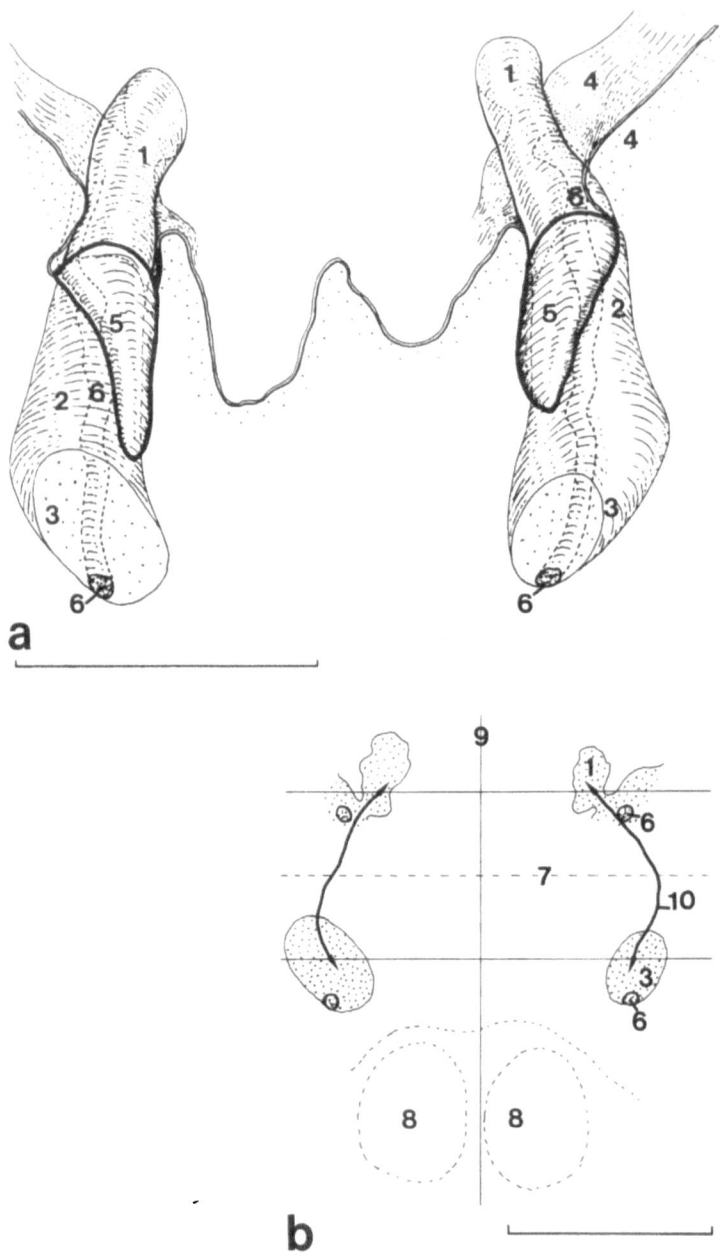

Fig. 23a,b. Reconstruction of the gubernaculum of a 29-mm CRL female embryo. Same embryo as in Fig. 22. **a** Total view of gubernaculum from ventral side and position of processus vaginalis. **b** Diagram shows the course of the gubernaculum. *1,* abdominal part of gubernaculum; *2,* interstitial part of gubernaculum; *3,* subcutaneous part of gubernaculum; *4,* abdominal wall; *5,* processus vaginalis; *6,* genital branch of genitofemoral nerve; *7,* entrance of gubernaculum into the abdominal wall; *8,* pubic bones; *9,* median plane; *10,* course of gubernaculum. *Bars,* 1 mm

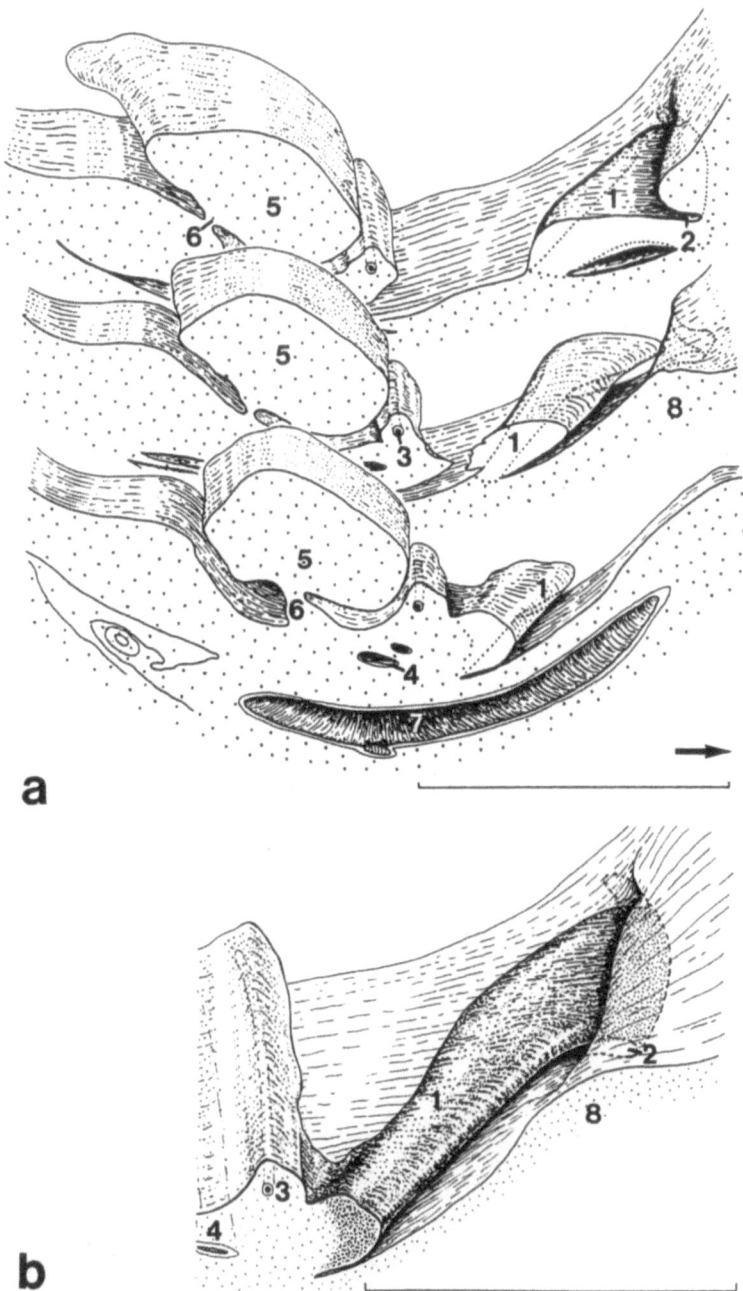

Fig. 24a,b. Plastic reconstruction of the whole left abdominal gubernaculum (**a**) and tripartite reconstruction based on sagittal sections of a 32-mm CRL male embryo (**b**). Medial view. *1*, abdominal gubernaculum; *2*, processus vaginalis; *3*, müllerian duct; *4*, wolffian duct; *5*, testis; *6*, mesorchium; *7*, left umbilical artery; *8*, abdominal wall. *Bars*, 1 mm

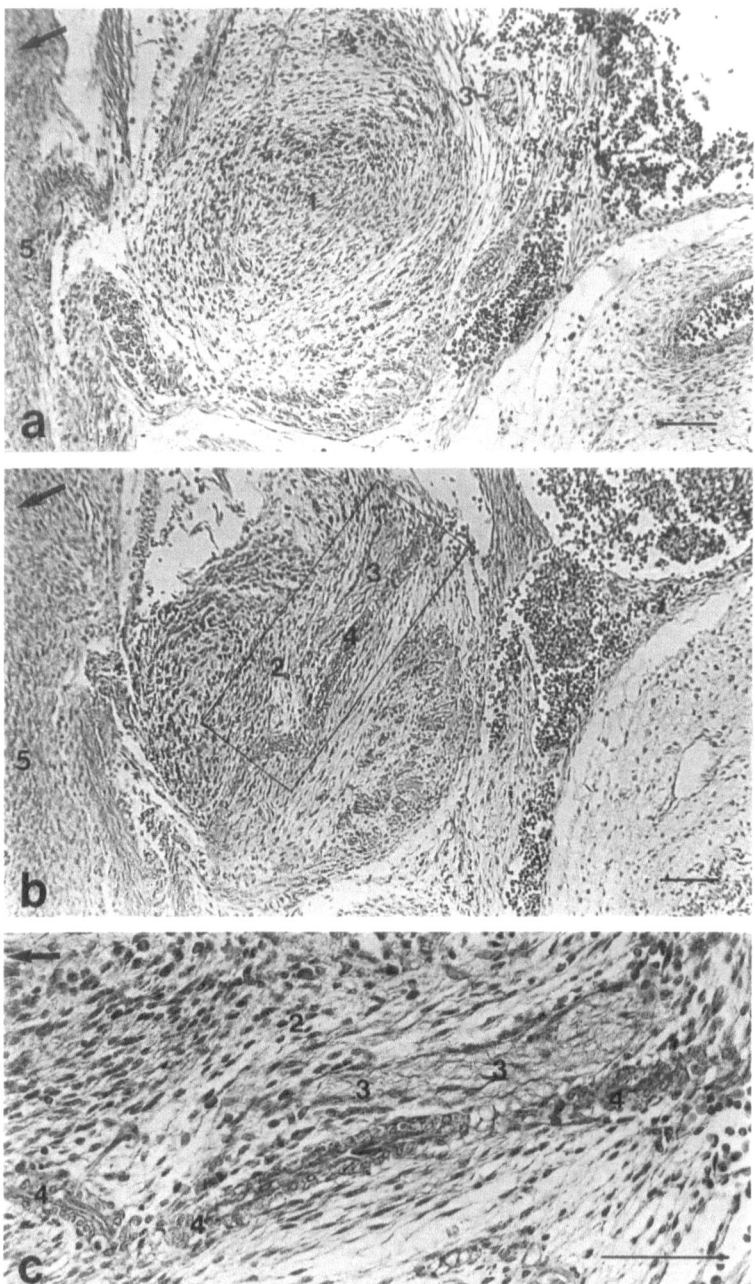

Fig. 25a–c. Transverse sections through the gubernaculum of a 38-mm CRL female embryo. **a** Section through the right interstitial gubernaculum (round ligament of uterus) with cross-sectioned genital branch of the genitofemoral nerve. **b** More distal section than in **a**, with subcutaneous ligament and longitudinal sectioned nerve and artery. **c** Higher magnification of *rectangle* in **b**. *1*, interstitial part of gubernaculum; *2*, subcutaneous part of gubernaculum; *3*, genital branch of genitofemoral nerve; *4*, arteria ligamenti teretis uteri; *5*, abdominal wall. *Bars,* 0.1 mm

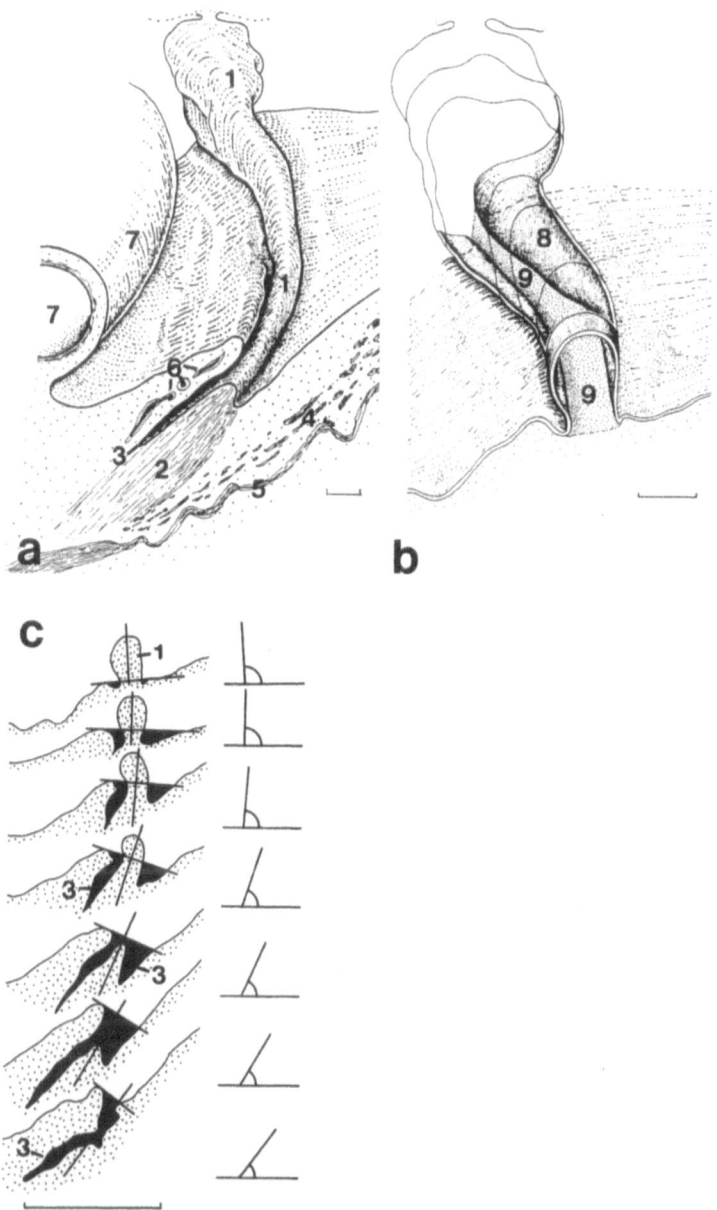

Fig. 26a–c. Graphic reconstruction of the left abdominal gubernaculum of a 45-mm CRL male foetus, sectioned frontally. **a** Total view of the gubernaculum. **b** Peritoneal coat and attachment zone of gubernaculum with abdominal wall. **c** *Left side,* axis of gubernaculum and peritoneal recessus forming the processus vaginalis peritonei. *Right side,* angle of gubernaculum with a transverse plane. *1,* abdominal gubernaculum; *2,* interstitial gubernaculum; *3,* processus vaginalis peritonei; *4,* abdominal muscles; *5,* aponeurosis of external oblique muscle; *6,* left inferior epigastric vessels; *7,* left umbilical artery; *8,* peritoneal coat; *9,* adhesion zone of abdominal wall and gubernaculum. *Bars* **a, b,** 0.1 mm; **c,** 0.5 mm

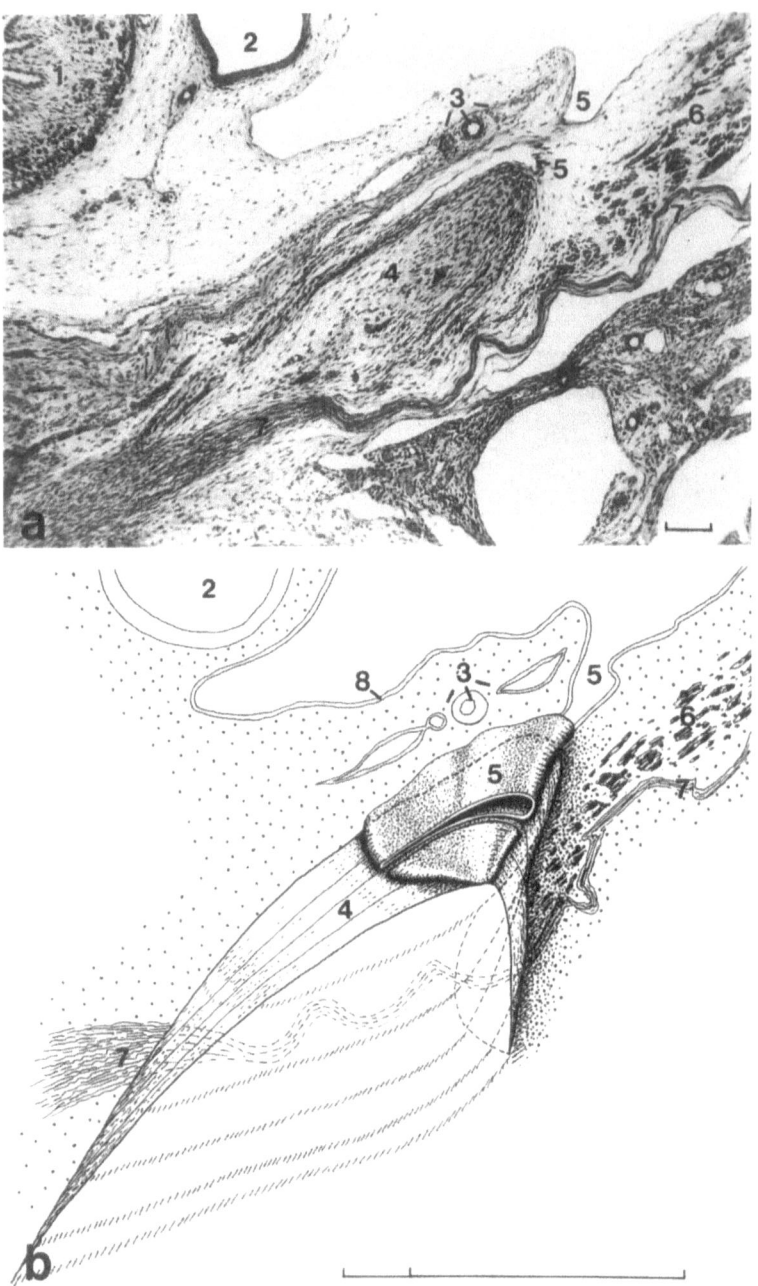

Fig. 27. a Frontal section through the same foetus as in Fig. 26 with the interstitial part of the left gubernaculum. **b** Partial reconstruction of gubernaculum and processus vaginalis. *1*, urinary bladder; *2*, left umbilical artery; *3*, left inferior epigastric vessels; *4*, interstitial part of gubernaculum; *5*, processus vaginalis; *6*, abdominal muscles; *7*, aponeurosis of external abdominal muscle; *8*, parietal peritoneum. *Bars* **a**, 0.1 mm; **b**, 0.5 mm

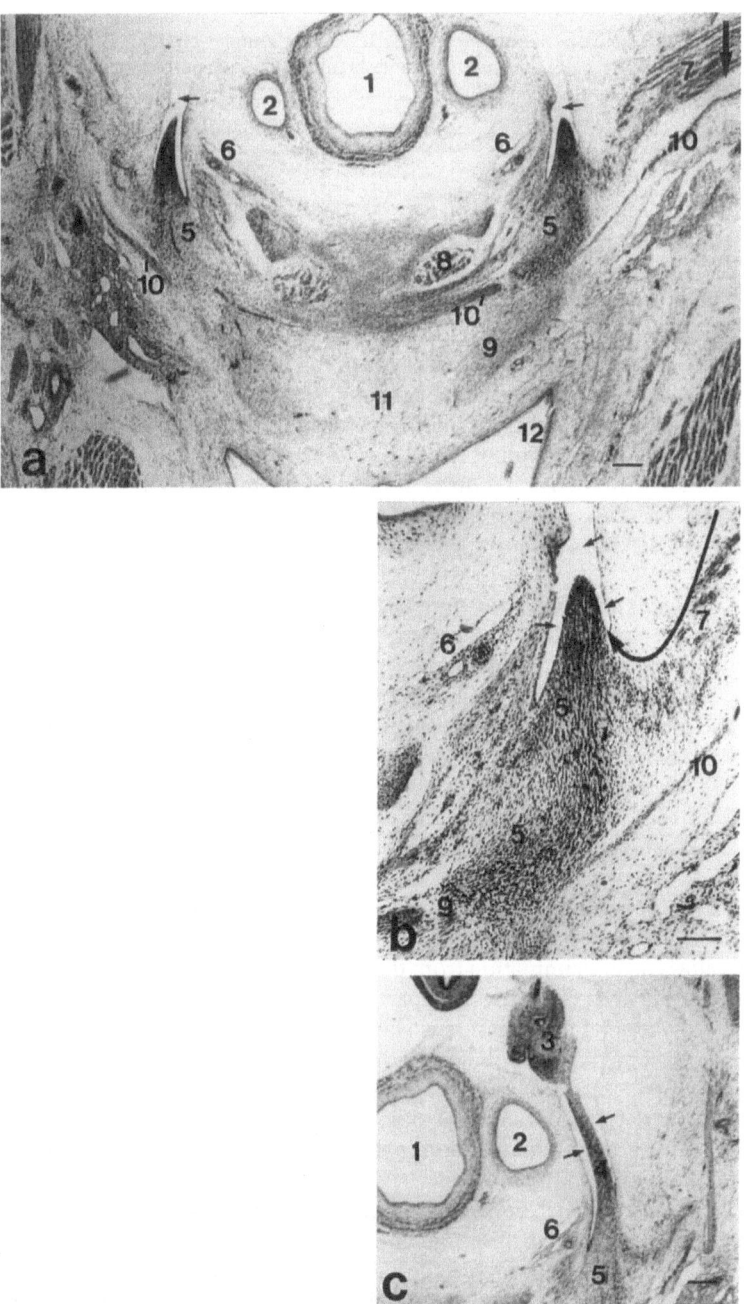

Fig. 28a–c. Transverse sections of another 45-mm CRL male foetus. **a** Overall view showing both interstitial gubernacula with pubic region. **b** Detail of the left gubernaculum in **a**. Note the reflection of inner abdominal muscles shown by the curved arrow. **c** Some sections more cranial than in **a**. Note the deep processus vaginalis at the medial side of the gubernaculum. *1*, urinary bladder; *2*, umbilical arteries; *3*, mesonephros with genital ducts; *4*, abdominal part of gubernaculum; *5*, interstitial part of

descent is not synchronous at both sides, one side is still in phase IIa. At the side of advanced development, the free caudal pole of the right testis shows up ventrally and overgrows the mesonephros and genital ducts. The cranial mesonephric ligament inserts near the cranial pole of testis and the caudal pole of the testis is connected by a ligament with the dorsal mesenchymal layer of genital ducts. This structure has the same position as the anlage of the ovarian ligament.

The prominent abdominal gubernaculum inserts more ventrally at the crossing point of the ducts. After preparation of the testis (Fig. 31a), the fixation of its caudal part is seen in the angle of the genital ducts at their dorsal mesenchyme, opposite to the gubernaculum. The anlage of the epididymis is found on the dorsal side of the testes (Fig. 31b) and seems to lie intraperitoneally. The gubernaculum is still a thin cord.

When comparing this stage with a 70-mm CRL male embryo (Fig. 32a), the growth of the testis is conspicuous to the other organs. It expands ventrocaudally, thus overriding genital ducts and approaches the inner inguinal ring. The caudal poles of both testes are free, the insertion of the still thin gubernaculum is the same as before (Fig. 31b). At the medial side of the gubernaculum, a deep inguinal evagination of peritoneal cavity can be observed (Fig. 32b,c).

The overriding position of the testis becomes more obvious in Fig. 33. After preparation of the testis, the detail again shows that the testis has no contact to the gubernaculum (Fig. 33), but its most caudal fixation is at the dorsal mesenchyme of the genital ducts.

The insertion of the abdominal gubernaculum at the duct mesenchyme can be concluded from the serial reconstruction (Fig. 34). The cell-rich structure of the gubernaculum allows its discrimination from the loosely arranged duct mesenchyme.

The relationship between the interstitial part of the gubernaculum, the processus vaginalis and the abdominal musculature is shown in Fig. 35a–c. The peritoneum is firmly attached ventromedially to the interstitial gubernaculum. The outgrowing gubernaculum with adhering processus vaginalis peritonei penetrates into the abdominal wall (Fig. 35b). At its dorsolateral side, the gubernaculum is surrounded by the aponeurosis of external oblique abdominal muscle and muscle fibres deriving from the inner oblique abdominal muscles.

In contrast to the testes which reveal a craniocaudal axis and whose caudal poles override the genital ducts, the position of the ovaries is quite different. In Fig. 36, photographed from a 78-mm CRL foetus, the caudal convergence of the ovary axis is obvious. The differentiating müllerian ducts build up a caudal barrier. The inferior pole of the ovary points dorsomedially and does not override the genital ducts. It is attached with the anlage of the ligamentum ovarii proprium at the dorsal side of the superior angle (Tubenwinkel) of the developing uterus. Opposite to it, at the ventral side of the genital ducts, the prominent gubernaculum (round ligament) runs semicircularly to the inner inguinal ring.

Fig. 29. a Transverse section through a female foetus slightly smaller than 50 mm CRL.

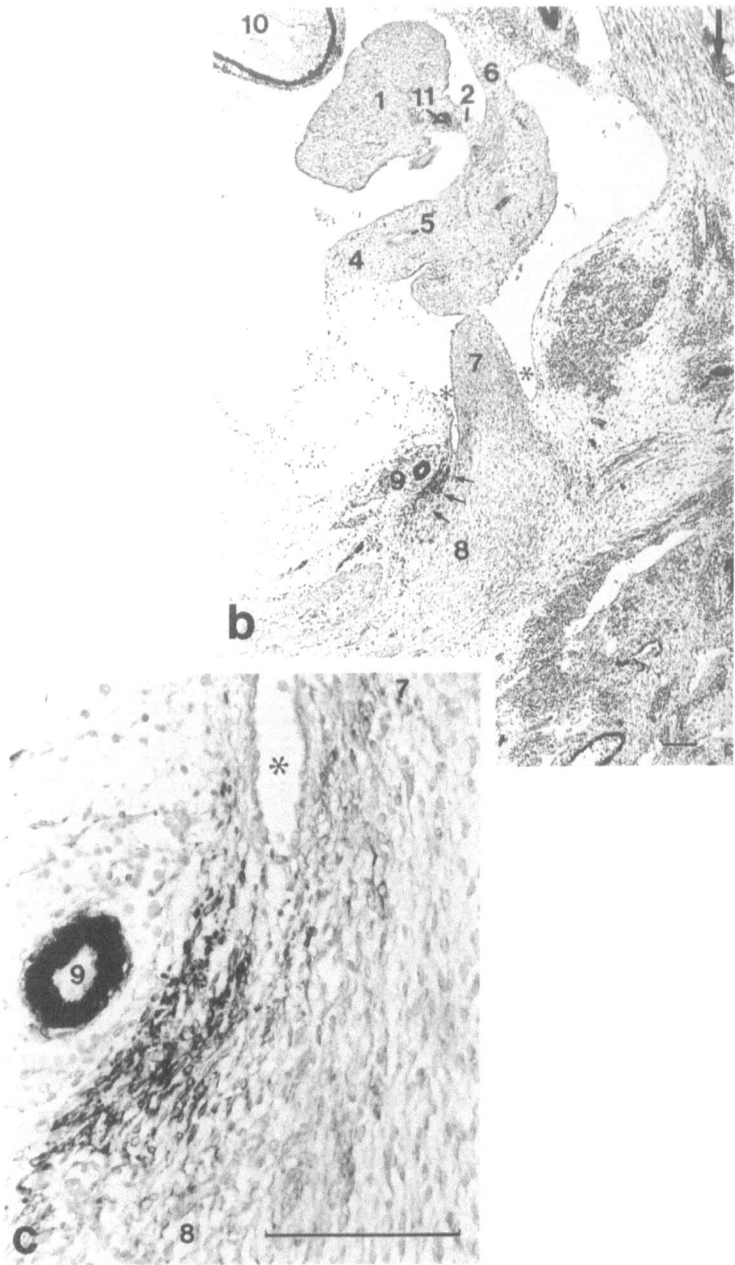

Fig. 29. b-c, b Immunostaining with anti-smooth muscle actin. Note heavy staining of the arterial wall. **c** Detail of **b**. *1*, left ovary; *2*, mesovarium; *3*, dorsal superior angle (Tubenwinkel) of uterus with anlage of ligamentum ovarii proprium (circle), *4*, anlage of fundus uteri; *5*, müllerian duct; *6*, radix of broad ligament; *7*, abdominal part of gubernaculum/anlage of round ligament of uterus; *8*, interstitial part of gubernaculum; *9*, left inferior epigastric vessels; *10*, intestine; *11*, ovarian artery. The *asterisk* shows the processus vaginalis, *arrows* show smooth muscles in the gubernaculum. *Bars*, 0.1 mm

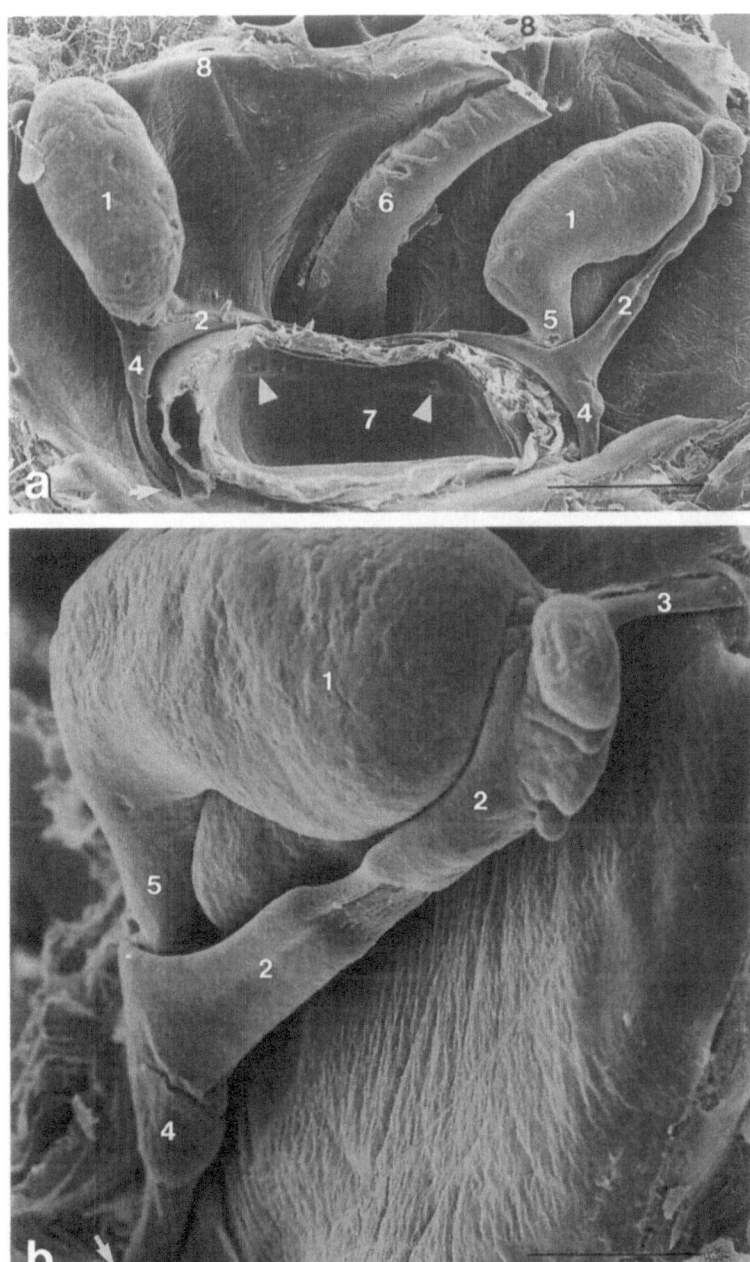

Fig. 30a,b. SEM micrographs of a 55-mm CRL male foetus. **a** ventrocranial view of the abdomen. Folds of the intestine and part of the urinary bladder are removed. *Bar,* 1 mm. **b** Magnification of the left testis, genital ducts and abdominal gubernaculum. Cranial view. *Bar,* 0.5 mm. *1,* testis; *2,* genital ducts; *3* cranial mesonephric ligament; *4,* abdominal gubernaculum; *5,* connection of caudal pole of testis with the mesenchyme of genital ducts; *6,* rectum; *7,* urinary bladder; *8,* ureters. The *arrow* shows the inner inguinal ring, *arrowheads* show openings of ureters

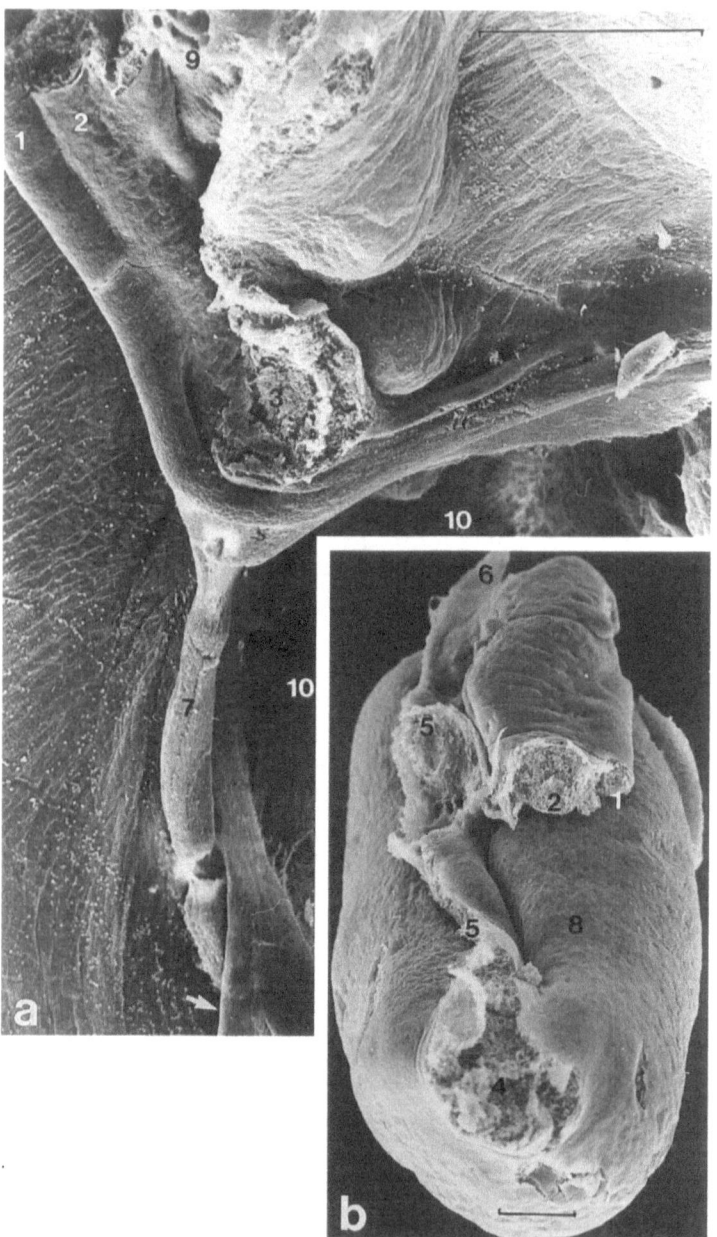

Fig. 31a,b. SEM micrograph of the same foetus as in Fig. 30 but after removal of the testis. a
Magnification of the attachment (zone of fracture) of the caudal part of the right testis dorsal to both
ducts. Cranial view. *Bar,* 0.5 mm **b** Dorsocranial view of removed testis. *Bar,* 0.1 mm. *1,* müllerian
duct; *2,* wolffian duct; *3,* fracture zone of the caudal part of testis at mesenchyme of genital ducts; *4*
opposite side at isolated testis; *5,* ligament which connects testis and epididymis with the posterior
abdominal wall (mesorchium); *6,* cranial mesonephric ligament; *7,* right abdominal gubernaculum
8, testis; *9,* epididymis (mesonephric rudiment); *10,* urinary bladder. The *arrow* shows the interna
inguinal ring

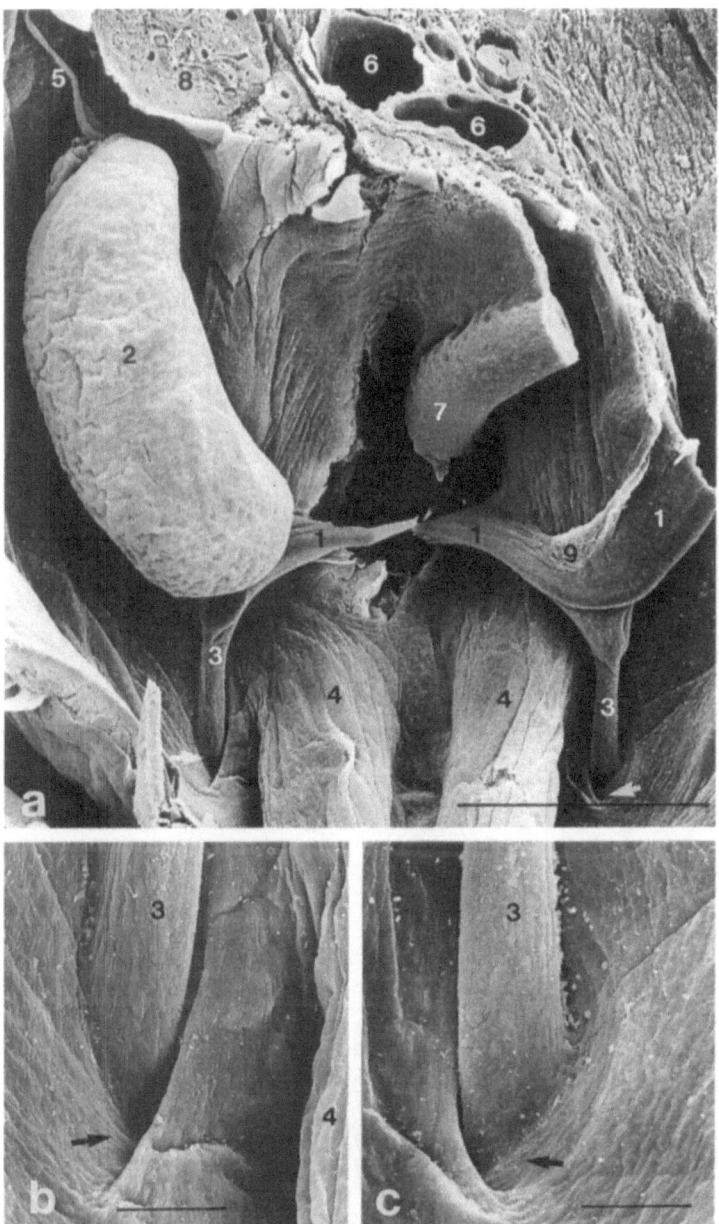

Fig. 32a,b. SEM micrograph from the dissected caudal half of body with opened abdomen of a 70 mm CRL male foetus. Left testis, urinary bladder, and folding of intestine removed. Ventral view. **a** Overall view. *Bar* 1 mm **b** and **c** Details with right and left inner inguinal rings. *Bars* 0.1 mm. *1* Müllerian and Wolffian ducts, *2* right testis, *3* abdominal gubernaculum, *4* umbilical arteries, *5* cranial mesonephric ligament, *6* aorta and inferior vena cava, *7* rectum, *8* transsectioned metanephros, *9* caudal attachment of testis (zone of fracture at the mesenchyme of genital ducts), *arrows* inner inguinal rings

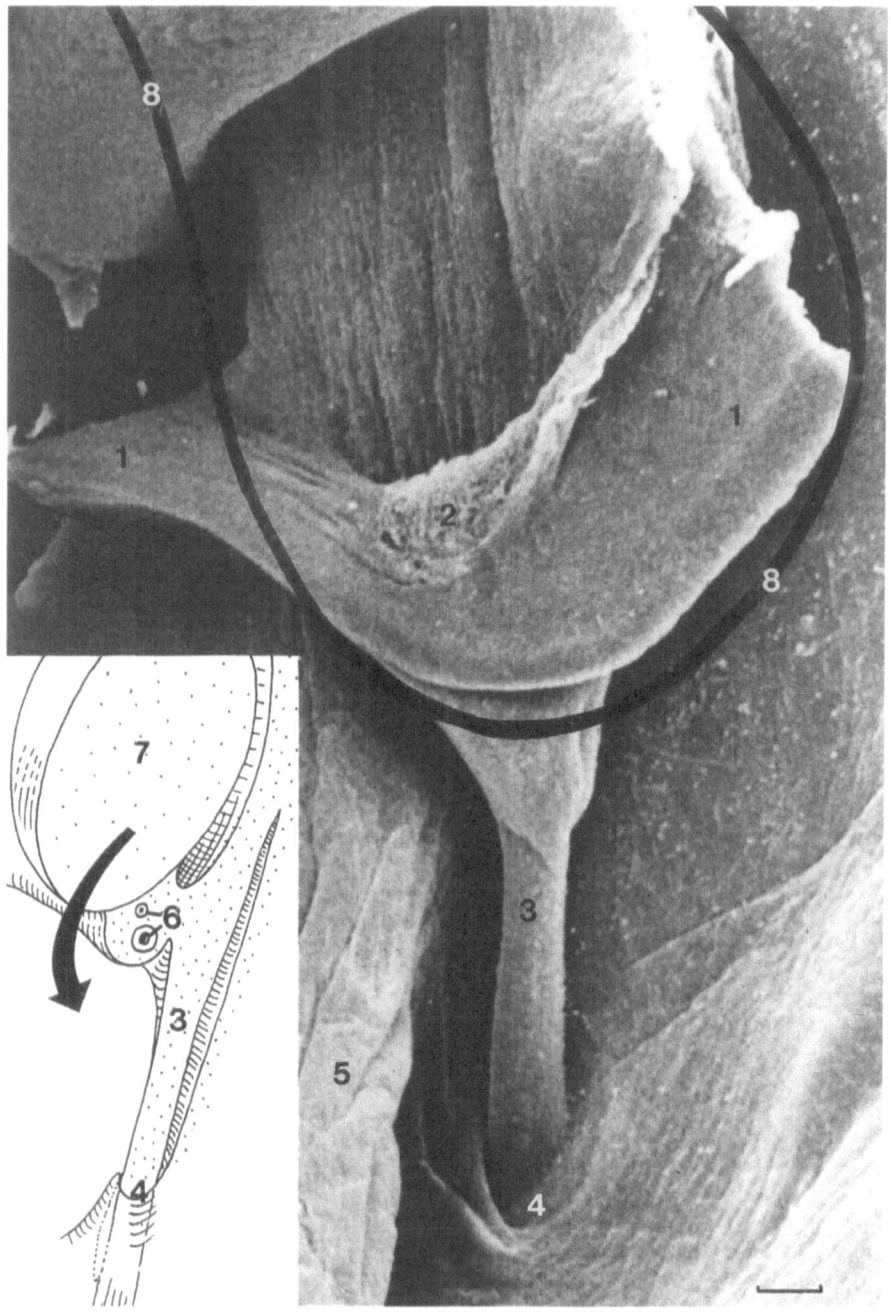

Fig. 33. Detail of Fig. 32 showing the attachment of the caudal part of the testis and illustrating the ventrocaudal movement of testis (*diagram*). The *black line* indicates the outlines of the removed testis. *1*, peritoneal surface of mesenchyme of genital ducts; *2*, attachment zone (fractured) at duct mesenchyme; *3* abdominal gubernaculum; *4*, internal inguinal ring; *5*, left umbilical artery; *6*, müllerian and wolffian duct; *7*, testis; *8*, outline of removed testis. *Bar*, 0.1 mm

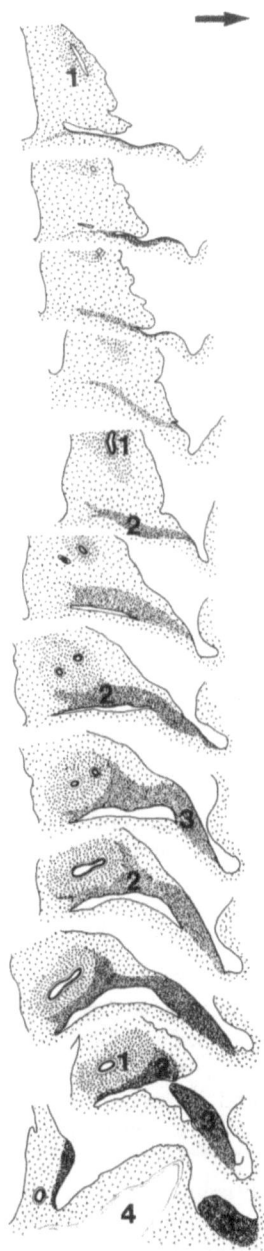

Fig. 34. Series of drawings demonstrating course and insertion of the gubernaculum into the mesenchymal layer of both genital ducts of a 70-mm CRL male foetus from medial (*bottom*) to lateral (*top*) side. *1*, wolffian duct; *2*, gubernacular cells inserting the duct mesenchyme; *3*, abdominal gubernaculum; *4*, left umbilical artery. *Bar*, 1 mm

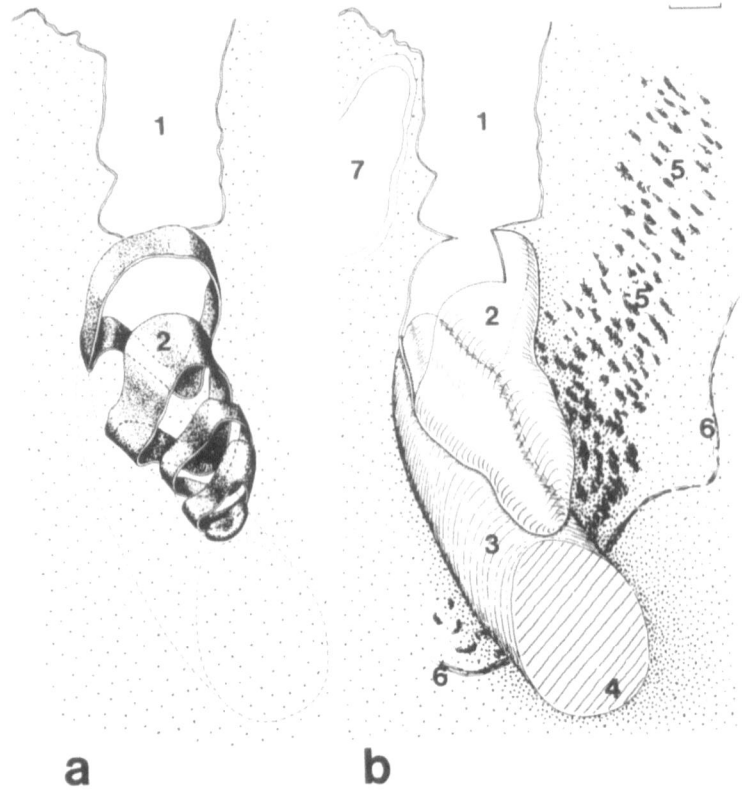

Fig. 35a–d. Left side of a 70-mm CRL male foetus. **a,b** Partial reconstruction of the interstitial gubernaculum with accompanying processus vaginalis which are shown as relief projecting over the frontal section of the abdominal wall.

Meanwhile the sex-specific differentiation of external genitalia has begun. In the 84-mm CRL male foetus (Fig. 37), the genital swellings have fused and the scrotum is formed. The latter consists of two chambers filled with loose mesenchyme. It does not contain smooth muscle cells, except in the wall of arteries and the anlage of the tunica dartos. Yet, we cannot find traces of the gubernaculum inserting into the bottom of the scrotum.

The gubernaculum does not yet differ in both sexes. It is build up by a cell-rich connective tissue containing smooth muscle cells as revealed by immunohistochemistry (Fig. 38c,d). Its lateral border is marked by a sheath of skeletal musculature (Fig. 38a,b).

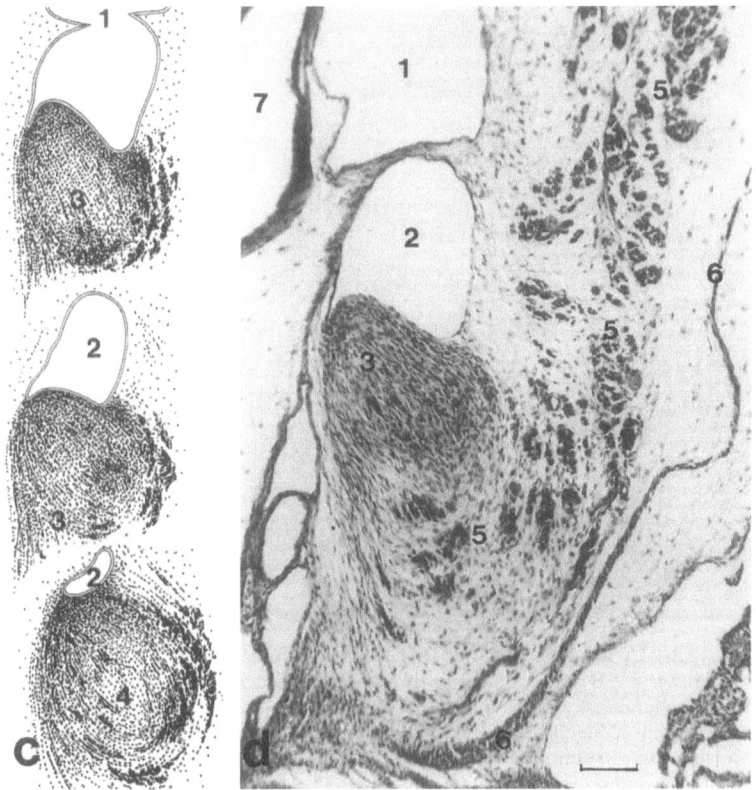

Fig. 35 c—d. c Drawings of serial sections. **d** Original frontal histological section. *1*, abdominal processus vaginalis peritonei; *2*, interstitial processus vaginalis and its cavity; *3*, interstitial part of gubernaculum; *4*, transition into subcutaneous part of gubernaculum; *5*, inner abdominal muscles; *6*, aponeurosis of external abdominal muscle; *7*, left umbilical artery. *Bars*, 0.1 mm

4.1.5
Phase IIIa:
Testis Glides over Genital Ducts

An important progress in testicular descent occurs when the caudal pole of the testis glides over both ducts and for the first time direct contact between the testes and the gubernacula is established. This movement is closely related to the involution of the müllerian duct. This dislocation permits the testis as well as the wolffian duct (later cauda epididymidis) to dip into the swollen, jelly-like mass of the abdominal gubernaculum and melt with it. Henceforward, the epididymidi and ductus deferens are in their definitive position, that is dorsolateral to the testis as shown in the 120-mm CRL male foetus (Fig. 39a). The graphic reconstruction of the abdominal gubernaculum caudal to the testis demonstrates that the cauda epididymidis is the most advanced structure in the descent of genital organs (Fig. 39b).

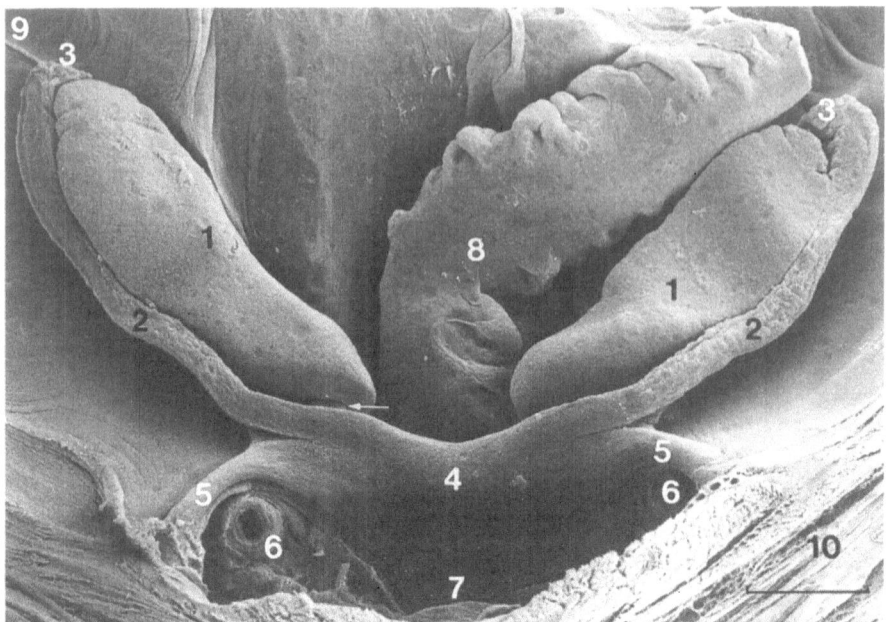

Fig. 36. SEM micrograph of relationships between the gonads and genital ducts in a 78-mm CRL female embryo. Ventral view. *1*, ovary; *2*, müllerian and wolffian duct; *3*, funnel region of fallopian tube; *4*, anlage of uterine fundus; *5*, insertion of round ligament of uterus; *6*, orifices of ureters into the bladder; *7*, side of removed bladder; *8*, rectum; *9*, cranial mesonephric ligament; *10*, abdominal muscles. The *arrow* shows ligamentum ovarii proprium at the dorsal side of the superior angle of the uterus (Tubenwinkel). *Bar,* 1 mm

The interstitial gubernaculum (Fig. 39c) is denser and richer in fibrils than the abdominal gubernaculum at this stage. Histological staining (Fig. 39d,e) and immunostaining with anti-smooth muscle actin (not shown) reveal a considerable proportion of smooth muscle cells within the connective tissue of the gubernaculum. Distally, the latter shows disintegration into a loose mesenchyme which disperses in the pubic region (Fig. 39f).

4.1.6
Phase IV:
Testes at Inner Inguinal Ring

In the 170-mm CRL male foetus (about 20 weeks) the enormous increase of diameter of the gubernaculum can be demonstrated impressively (Fig. 40a). It is nearly as broad as the testis which perches on the gubernaculum overriding it ventromedially (Fig. 41). The interstitial gubernaculum tapers distally ending as thin cord (subcutaneous part) at the pubic region (Fig. 42). The lower part of the abdominal and the interstitial gubernaculum (Fig. 40b) are incompletely surrounded by a sheath of skeletal muscles from which single fibres detach. As can be seen in Fig. 40d, even in the

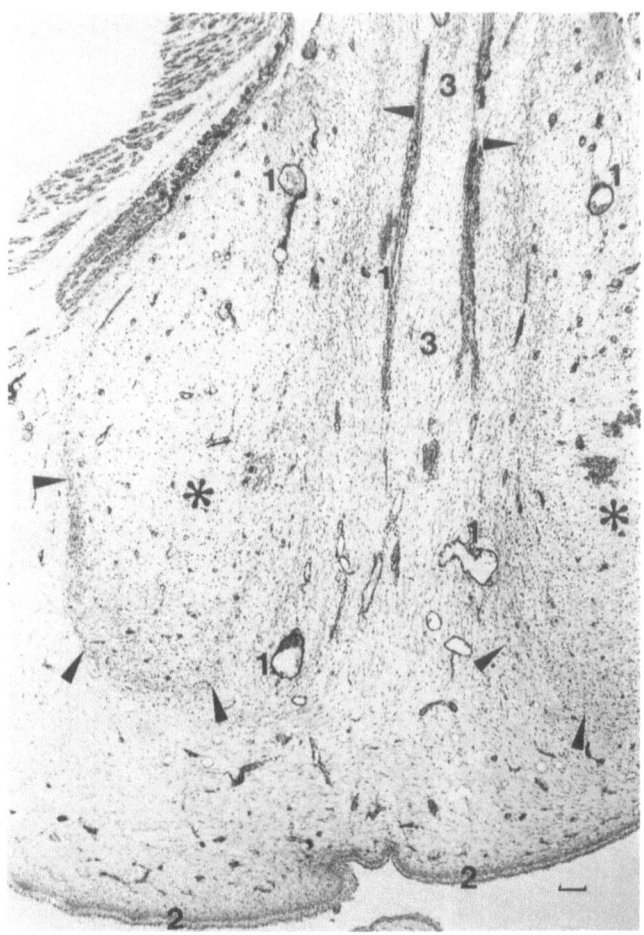

Fig. 37. Frontal section through the scrotum anlage of an 84-mm CRL embryo. Immunostaining with anti-smooth muscle actin. *1,* heavy reaction of the tunica media of the vessels; *2,* anlage of tunica dartos; *3,* anlage of scrotal septum. *Arrowheads* show the boundary of the scrotal chamber, *asterisks* indicate the scrotal chamber. *Bar,* 0.1 mm

middle of the gubernaculum are single fibres of skeletal musculature found. Furthermore, Azan and trichrome staining reveal a network of collagenous fibrils encircling the gubernaculum with a denser layer. The jelly-like core exhibits less fibrils.

The parallel course of the cremaster artery, the genital branch of genitofemoral nerve and the interstitial gubernaculum with direction to the pubic region becomes obvious in Fig. 42. The ramus genitalis is not only close to the gubernaculum but also precedes it into the scrotum. No connection exists between the gubernaculum with the bottom of the scrotum, which is already well developed. As a consequence of the gubernacular swelling, the inguinal channel widens.

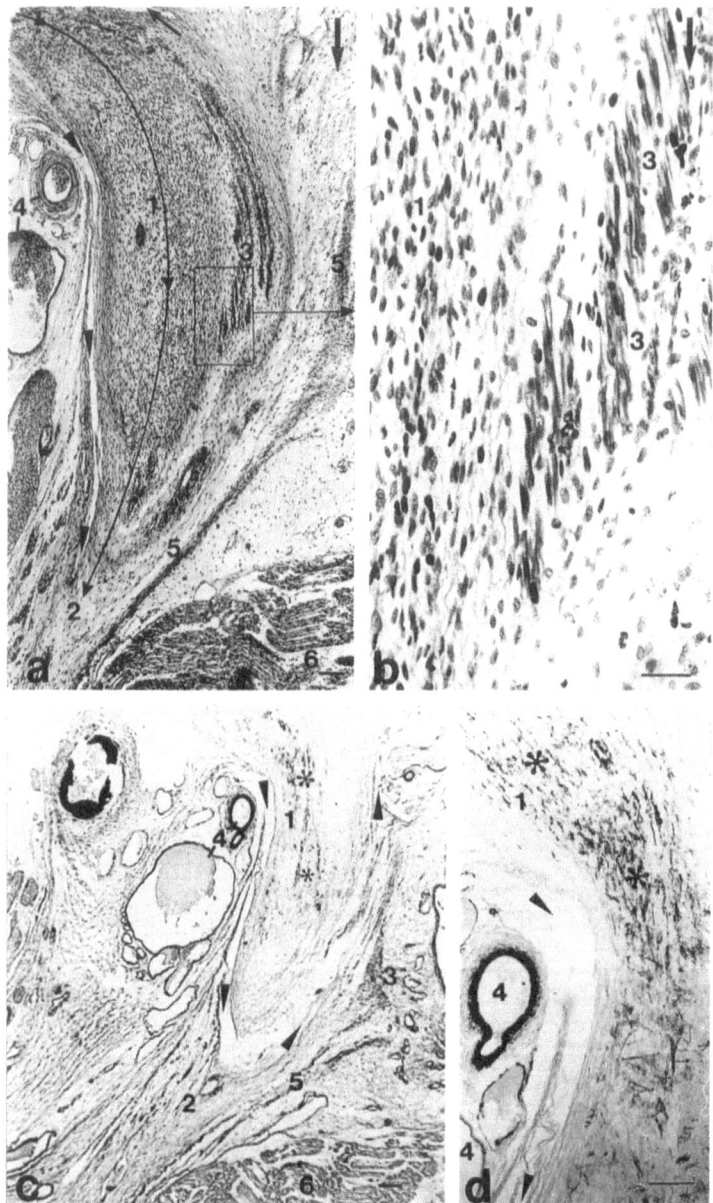

Fig. 38a–d. Transverse sections of a 105-mm CRL female foetus. **a** Overall view of interstitial gubernaculum. **b** Detail of *rectangle* in **a**. Note the skeletal musculature at the lateral margin of gubernaculum. **c,d** Immunolocalisation of smooth muscle actin. Note the heavy staining in the arterial wall and the network of smooth muscle cells in the proximal part of gubernaculum. *1*, interstitial gubernaculum (presumptive round ligament of uterus); *2*, subcutaneous part of gubernaculum; *3*, abdominal muscles in the border zone of gubernaculum; *4*, inferior epigastric vessels *5*, aponeurosis of external abdominal muscle *6*, thigh muscles. *Arrowheads* show the processus vaginalis peritonei, *black line with arrowheads* in **a** marks the arched course of gubernaculum, *asterisks* show localisation of smooth muscle actin. *Bars*, 0.1 mm

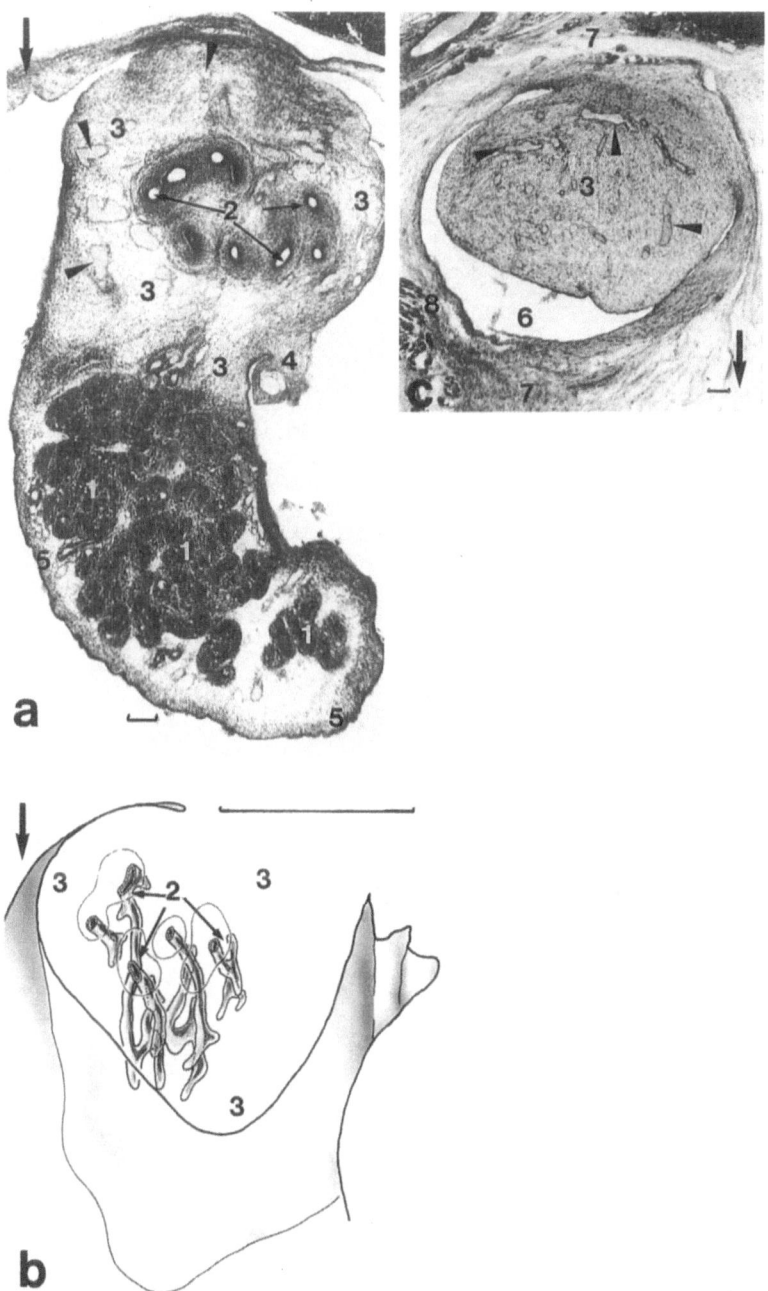

Fig. 39a–f. Transverse section through the right testis of a 12-cm CRL foetus. **a** Caudal part of the testis and epididymis protrude into the gubernacular mass. Testis has achieved its position ventral to epididymis and genital ducts. **b** Block reconstruction of a part of abdominal gubernaculum caudal to **a**, with the distal parts of cauda epididymidis. **c** Section through the interstitial gubernaculum caudal to **a** and **b**.

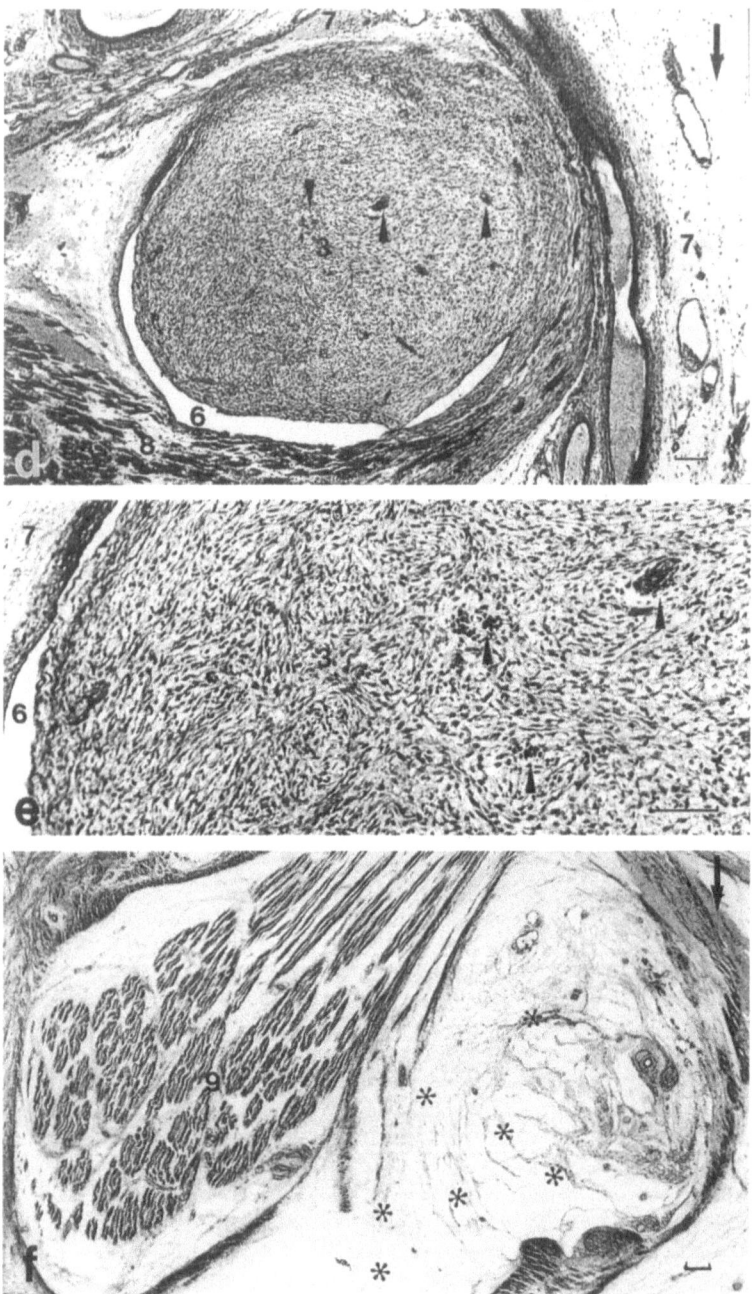

Fig. 39d–f. d Section slightly more distally. **e** Detail of gubernaculum with smooth muscle cells and fibrocytes. **f** Most distal part of the gubernaculum. *1*, testicular cords; *2*, ducts of epididymis; *3*, gubernaculum; *4*, remnant of wolffian duct/appendix epididymidis; *5*, tunica albuginea; *6*, processus vaginalis; *7*, abdominal wall; *8*, abdominal muscles; *9*, musculus pyramidalis. *Arrowheads* show blood vessels, *asterisks* show the fibrous distal part of the gubernaculum. *Bars* **a, c–f**, 0.1 mm; **b** 1 mm

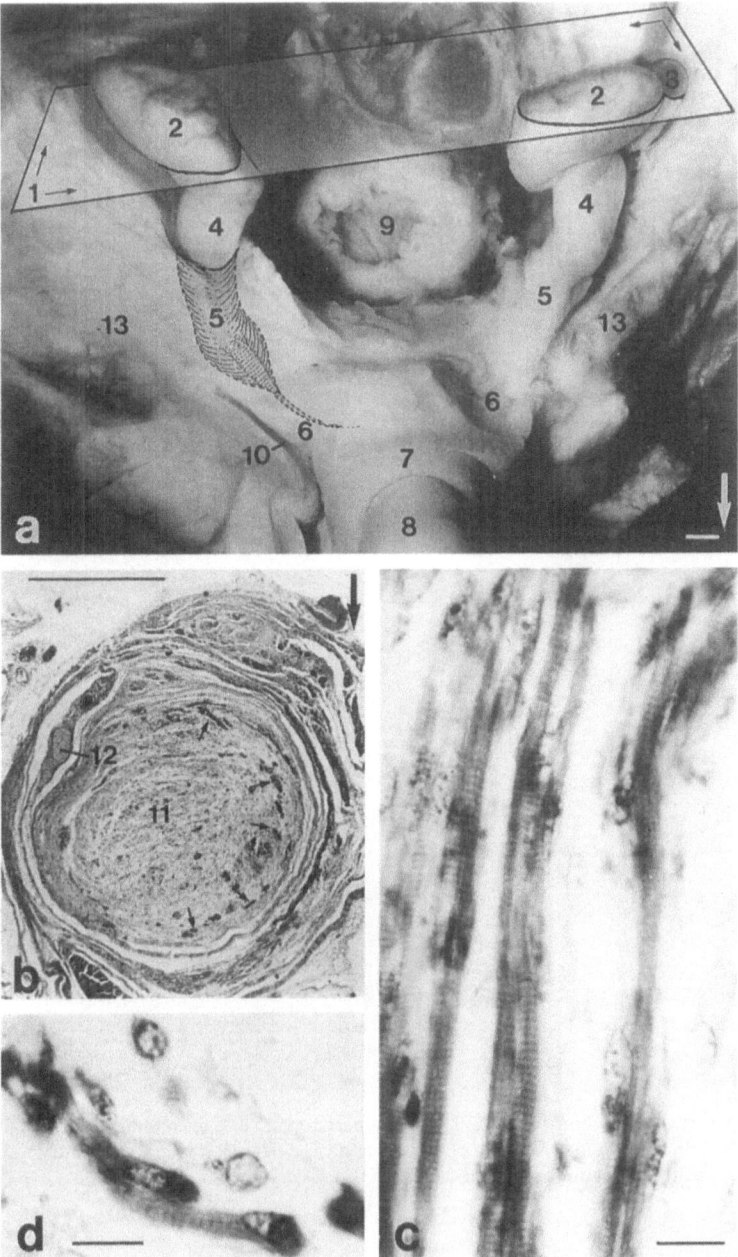

Fig. 40. a Preparation of the abdomen of a 17-cm CRL male foetus. Ventrocranial view to show testes and gubernacula. **b** Transverse section through right interstitial gubernaculum of the same foetus. **c** Detail of the skeletal musculature at the periphery of the gubernaculum and (**d**) in the centre of gubernaculum. *1*, plane of section for reconstruction (Fig. 42); *2*, testis; *3*, epididymis; *4*, abdominal gubernaculum; *5*, interstitial gubernaculum; *6*, subcutaneous gubernaculum; *7*, pubic region; *8*, penis; *9*, urinary bladder; *10*, inguinal bay; *11*, sectioned interstitial gubernaculum; *12*, genital branch of genitofemoral nerve; *13*, abdominal wall. *Small arrows* indicate striated musculature. *Bars*, 1 mm

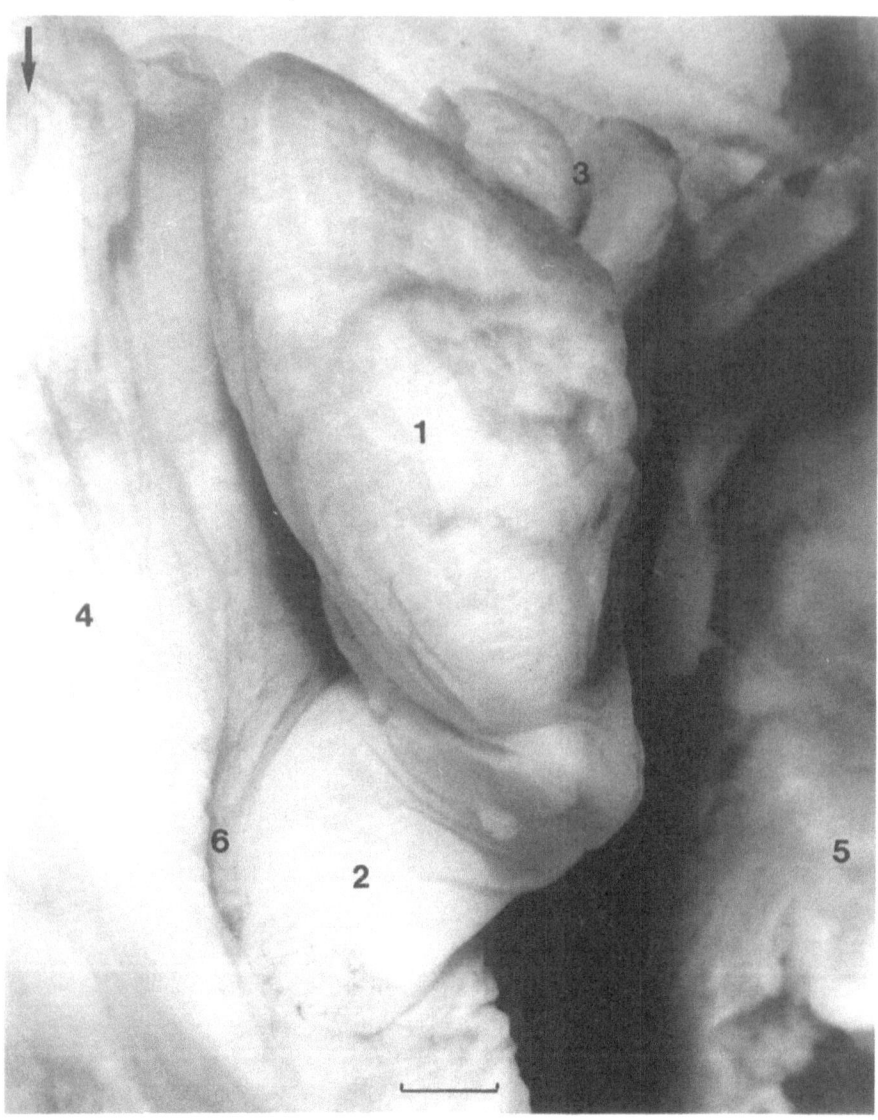

Fig. 41. Detail of Fig. 40a to show the dimensions of the gubernaculum and overriding testis. *1*, right testis; *2*, abdominal gubernaculum; *3*, head of epididymis and vas deferens; *4*, lateral abdominal wall; *5*, urinary bladder; *6*, internal inguinal ring. *Bar*, 1 mm

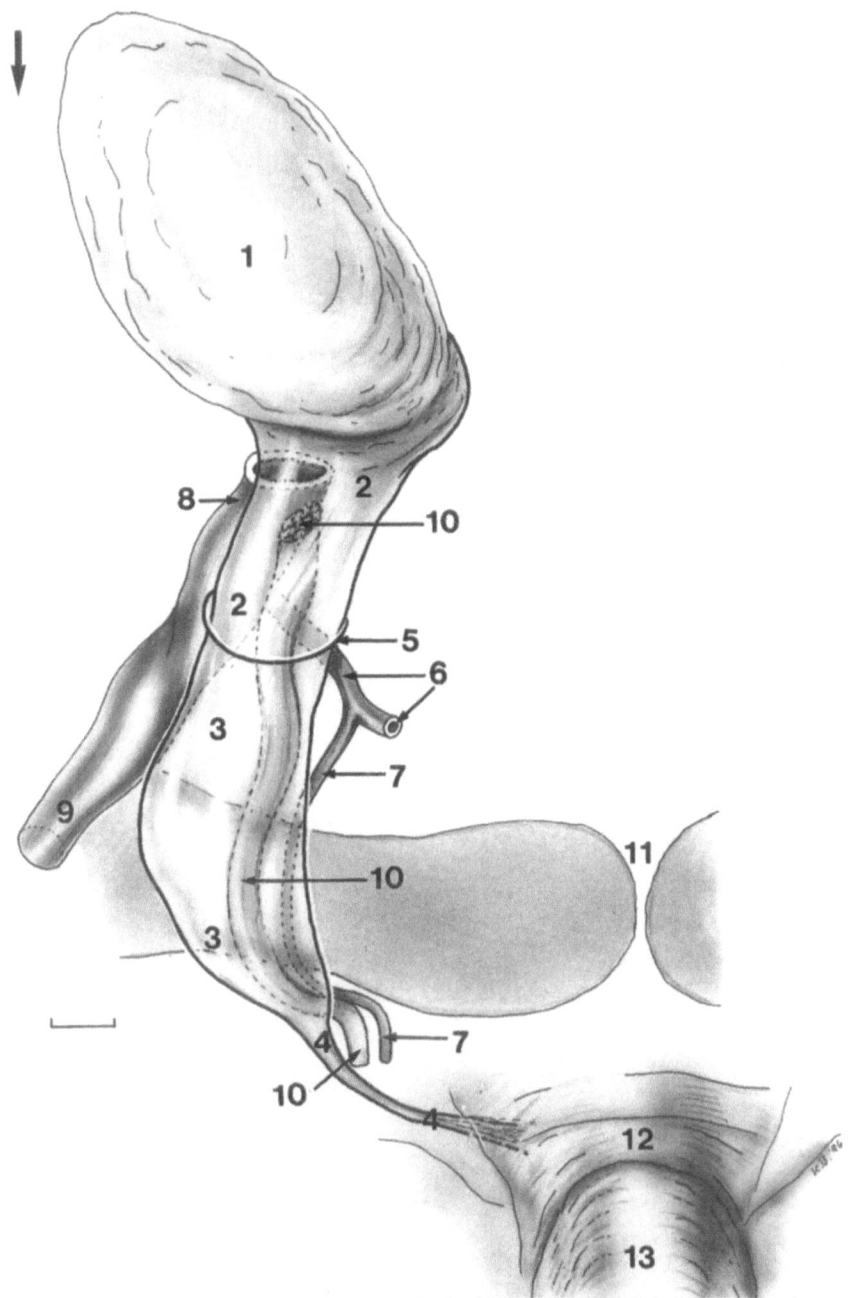

Fig. 42. Reconstruction of the whole right gubernaculum and surrounding structures of the same foetus as in Figs. 40, 41. Ventral view. *1*, testis; *2*, abdominal part; *3*, interstitial part; *4*, subcutaneous part of gubernaculum with distal mottling; *5*, internal inguinal ring; *6*, right inferior epigastric artery, *7*, right cremasteric artery; *8*, right external iliac artery; *9*, right femoral artery; *10*, genital branch of genitofemoral nerve; *11*, pubic symphysis; *12*, pubic region; *13*, penis. *Bar*, 1 mm

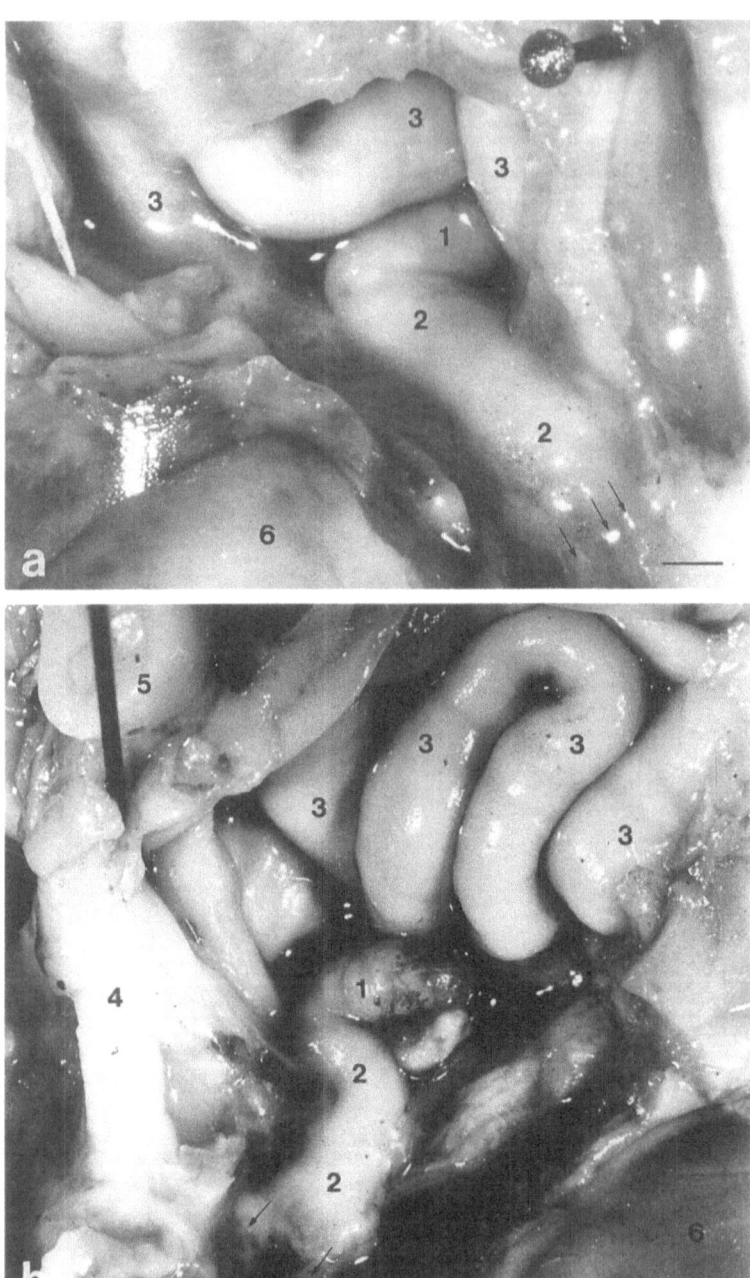

Fig. 43a,b. Photographs of the opened abdomen of a 25-cm CRL male foetus showing the testis in the abdomen and the gubernaculum in the prepared inguinal canal. **a** Right side. **b** Left side. *1*, testis; *2*, plug-like gubernaculum; *3*, intestine; *4*, penis; *5*, sectioned umbilical cord; *6*, thigh. *Arrows* show distal fraying of the gubernaculum still connected with the pubic region. *Bar*, 1 mm

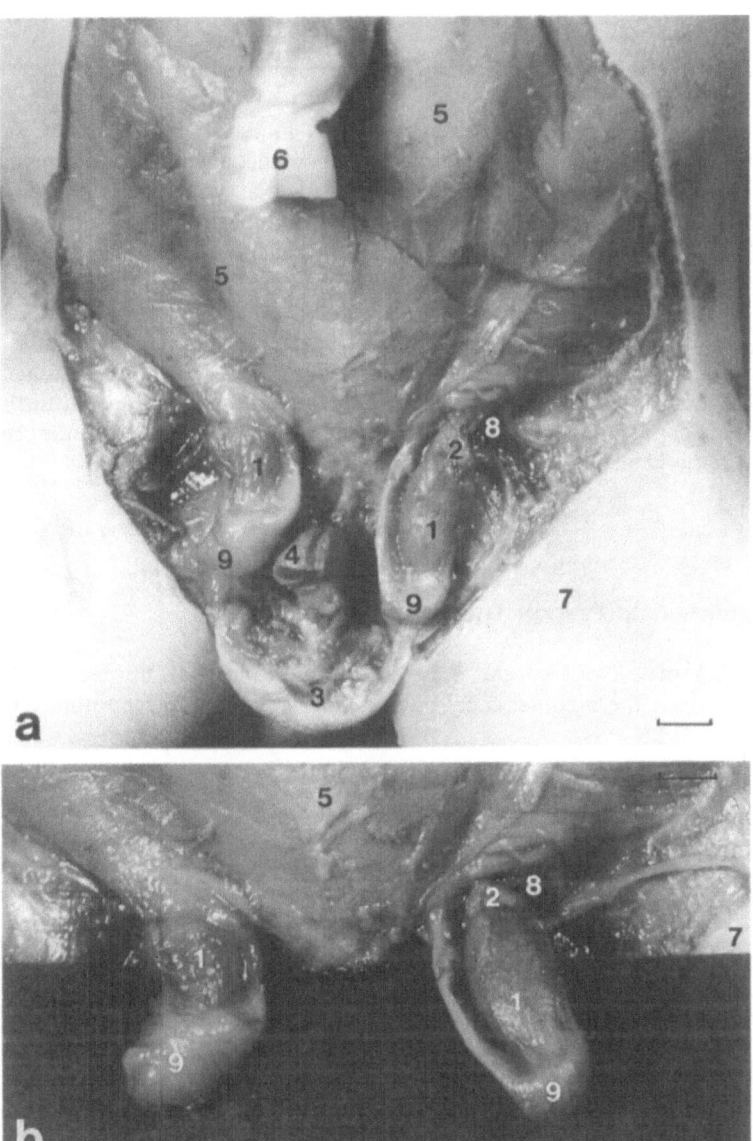

Fig. 44a,b. Photograph of the prepared testes of a 36-cm CRL foetus. Ventral view. a Gubernacula with coverings shortly before entrance into scrotum. Distally, no connection with pubic region visible, therefore, for demonstration (**b**), testes are photographed lying on cardboard. *1*, testes with coverings; *2*, head of epididymis with coverings; *3*, prepared scrotum; *4*, penis; *5*, abdominal wall *6*, umbilical cord; *7*, thigh; *8*, inguinal canal; *9*, covered gubernaculum. *Bars,* 2 mm

4.1.7
Phase V: Passage of the Testis and Epididymis Through the Inguinal Canal

This main phase of testicular descent commences in foetuses of 25–28 cm CRL. During this passage, the connection between the subcutaneous gubernaculum and pubic region loosens. The bolt-like gubernaculum is fixed at the caudal pole of the testis only proximally. Its structure is similar to the umbilical cord: According to Heyns (1987) the ratio of wet:dry mass of the gubernaculum increases and approaches that of the umbilical cord with the peak at 28 weeks. The high percentage of hyaluronate may be responsible for the increased water content (Heyns et al. 1990).

In our 25-cm CRL male embryo (about 5 months old), the testes are found near the internal inguinal ring still in the abdominal cavity (Fig. 43a,b). The plug-like gubernaculum lies in the inguinal canal. It disperses distally into the pubic region. A little bit later, these tender connections are stripped off. The immense volume of intestine convolutions is remarkable.

4.1.8
Phase Va:
Testes Have Finished the Passage Through the Inguinal Canal

In the 36-cm CRL male embryo (end of seventh month, Fig. 44), the testis with epididymis has passed the inguinal canal but has not yet entered the scrotum. The

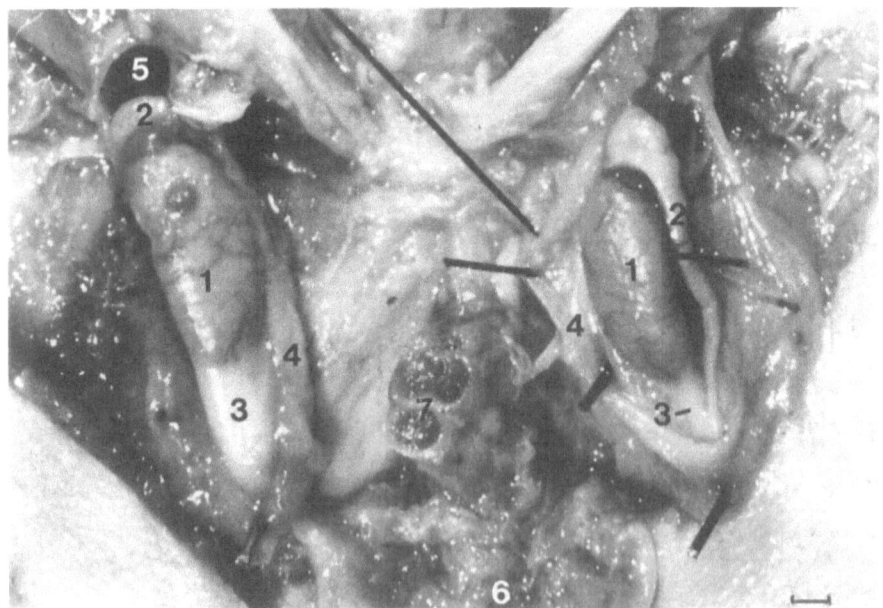

Fig. 45. Photograph of the testes and gubernacula of the same foetus as in Fig. 44. The processus vaginales peritonei are opened. *1*, testis; *2*, epididymis; *3*, gubernaculum; *4*, parietal layer of tunica vaginalis peritonei; *5*, canalis peritonei; *6*, scrotum *7*, penis. *Bar,* 3 mm

gubernaculum and both organs are covered by peritoneum of the prolonged processus vaginalis. No adhesion at the pubic region exists distally (Fig. 44b). The coverings of the testis can be shown as free blind sacs which have not yet received their definite position within the scrotum. After opening the peritoneal bag (Figs. 45, 46), the extremely shortened gubernaculum becomes visible. Since testicular descent is asymmetrical, the differentiation of the gubernaculum immediately after the passage through the inguinal canal can be demonstrated when comparing both sides (Fig. 46). At the retarded side (Fig. 46a) it has nearly the same diameter as the cone-shaped testis which is the leading end during descent while the head of the epididymis is the trailing end.

During and after the passage through the inguinal canal, the gubernaculum is shrinking. At the more advanced side (Fig. 46b), the testis is clearly bigger than the gubernaculum. The latter now consists of two strands which insert into the caudal poles of the testis or epididymis and fuse distally.

A proportional comparison of lengths and diameters of the gubernacula in extremely different developmental stages (embryo 22 mm CRL and foetus 360 mm CRL) gives an impression of the wide swelling of the gubernaculum. With a 16 times increase of CRL, the length increase of the gubernaculum is similar [13 times (0.91 mm–12 mm)], but the diameter of the gubernaculum increases by a factor of 55 (0.13 mm–7.2 mm).

4.1.9
Adult

One year after birth, the testes have reached their final scrotal position. In adults (Fig. 47), testes are covered by the visceral layer of the tunica vaginalis. Within the reflection of visceral into parietal layer, the remnant of the gubernaculum is found, known as scrotal ligament. In our material it measures 2.5 mm in length and 2 mm in diameter. It connects the tunica albuginea at the lower pole of testis with the internal spermatic fascia.

A synopsis of the phases of testicular descent is presented in Fig. 48:

Phase I: Contact of the mesonephric fold with the ventral abdominal wall.
Phase II: Forming of the abdominal gubernaculum and inner inguinal ring.
Phase IIa: Growth of the gubernaculum and the processus vaginalis peritonei. Ligamentous contact of the caudal pole of the testis with the dorsal mesenchyme of the genital ducts.
Phase III: Testis passes ventrad overriding the genital duct and swelling reaction of gubernaculum begins.
Phase IIIa: Testis and cauda epididymidis dip into the gubernacular mass.
Phase IV: Testis and epididymis in the gubernaculum. They approach the inner abdominal ring. Inguinal canal and external inguinal ring are formed.
Phase V: Passage of testis through the inguinal canal. Shrinkage of the gubernaculum begins.
Phase Va: Testis and epididymis migrate into scrotum.
Adult: Final position of testis in scrotum. The remnant of the gubernaculum is the scrotal ligament.

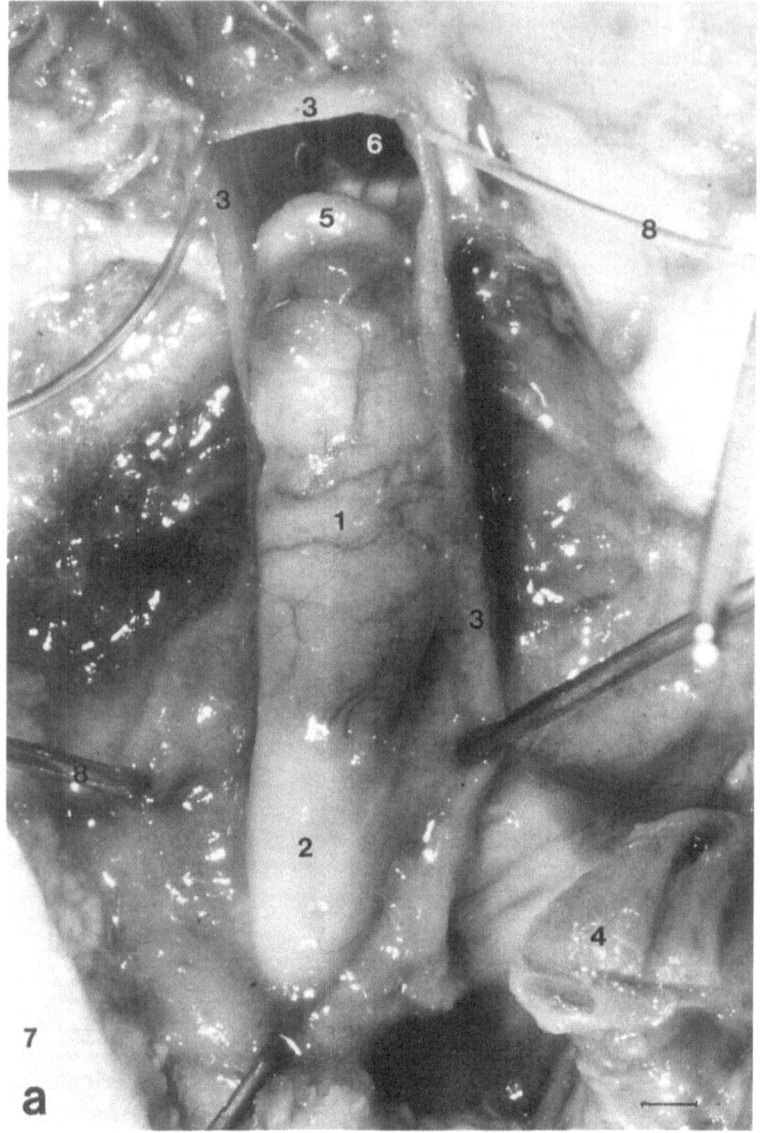

Fig. 46a. Details of Fig. 45. a Right testis with plug-like gubernaculum.

During these phases the orientation of the testes changes considerably (Fig. 49). They first appear as longitudinally oriented anlagen. They then have an almost transverse orientation, an arrangement which is characteristic for female gonads. During descent, the testes rotate into a more craniocaudal position. In adults they are again found parallel to the body axis.

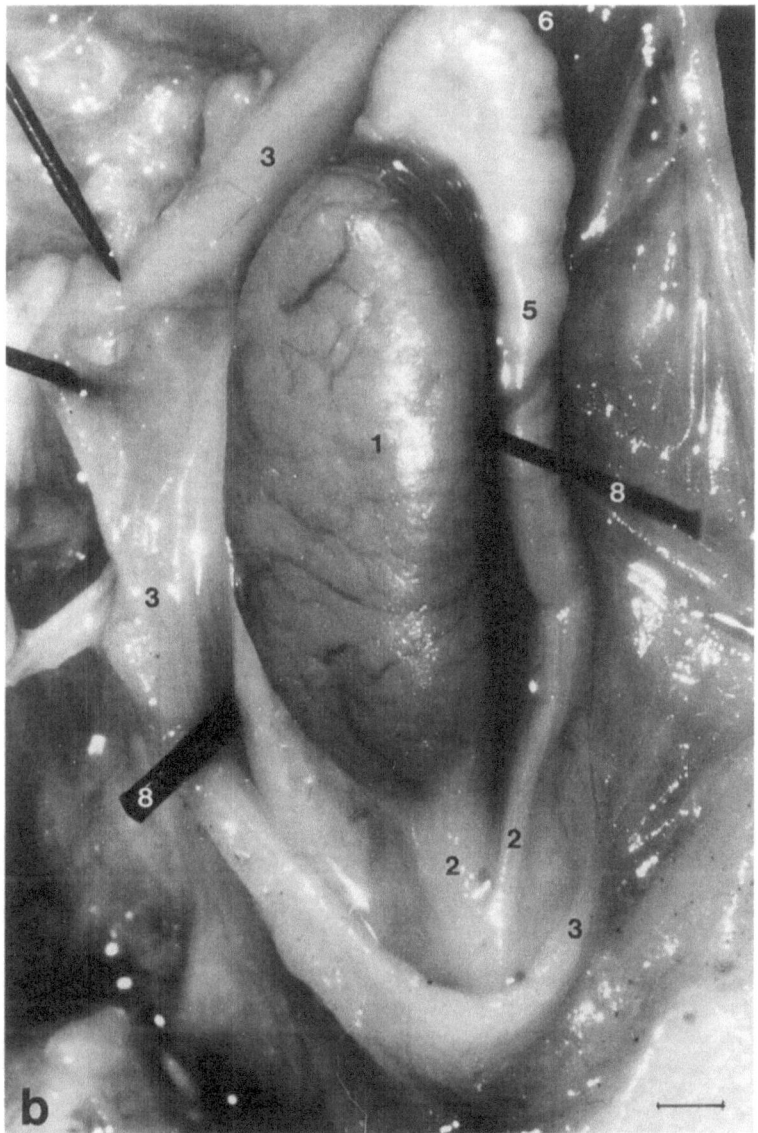

Fig 46 b. More advanced stage. left Testis and epididymis with two strands of gubernaculum joining distally to a single strand. *1*, testis; *2*, gubernaculum; *3*, coverings; *4*, penis, *5*, epididymis; *6*, canalis peritonei; *7*, thigh; *8*, needles and cords for preparation. *Bars*, 1 mm

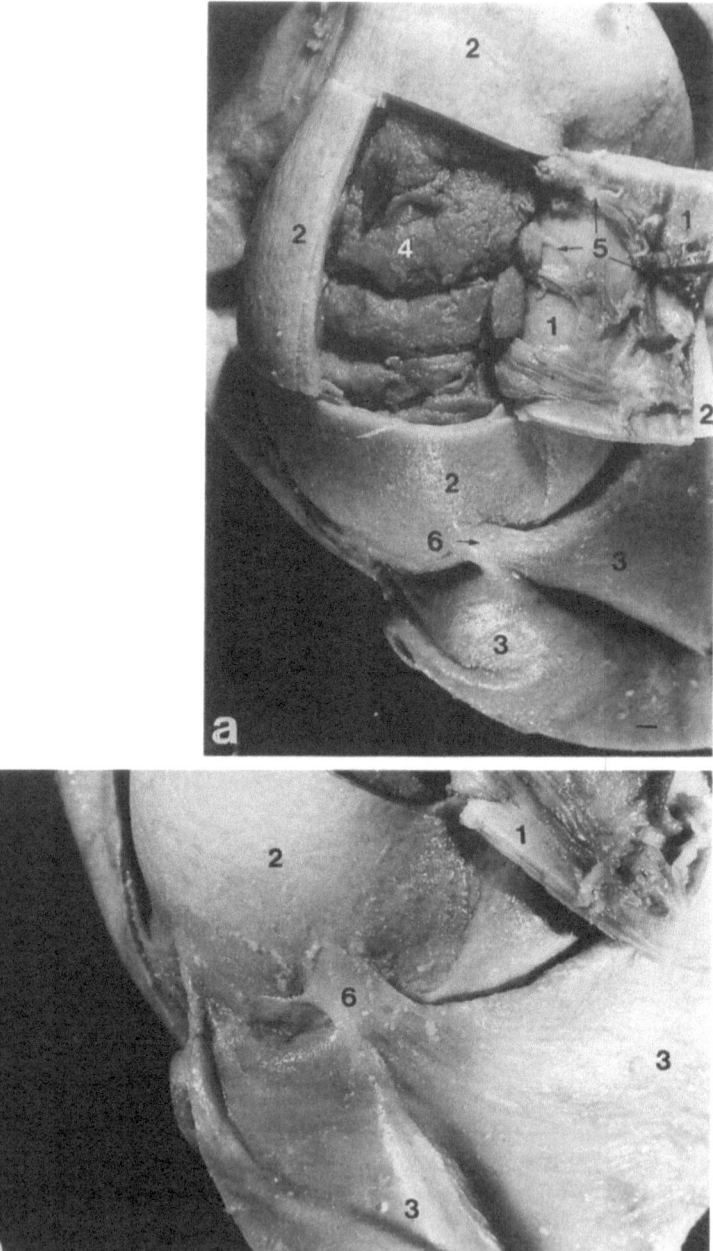

Fig. 47a,b. Adult left testis. **a** Fenestrated tunica albuginea and visceral layer of processus vaginalis peritonei. **b** Detail of **a**, with reflection of visceral into parietal layer of processus vaginalis peritonei. *1*, tunica albuginea; *2*, visceral layer (epiorchium); *3*, parietal layer (periorchium); *4*, parenchyma of testis; *5*, blood vessels; *6*, ligamentum scrotale testis (remnant of gubernaculum). *Bar*, 1 mm

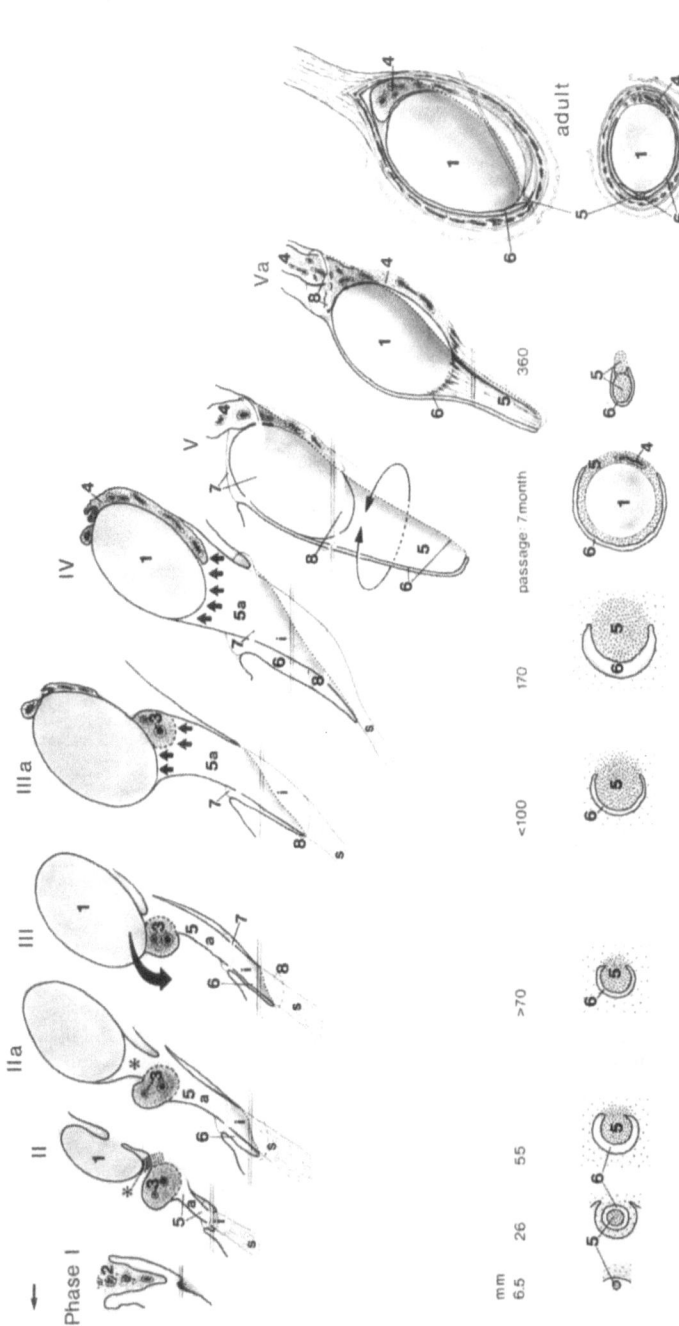

Fig. 48. Synopsis of testicular descent with phases I to adult. Sagittal view and transverse sections (*below*) in level of *double line*. *1*, gonad; *2*, mesonephros; *3*, wolffian and müllerian duct; *4*, Vas deferens/epididymis; *5*, gubernaculum (scrotal ligament); *6*, processus vaginalis/tunica vaginalis testis; *7*, internal inguinal ring; *8*, external inguinal ring. *Asterisks* show the link between the caudal pole of the testis and dorsal mesenchyme of genital ducts, *straight arrows* show ventral side, *curved arrow* shows direction of testicular migration, *short arrows* indicate attachment between the testis or epididymis and gubernaculum, *oval arrowline* indicates no connection between the bolt-like gubernaculum and neighbouring structures

63

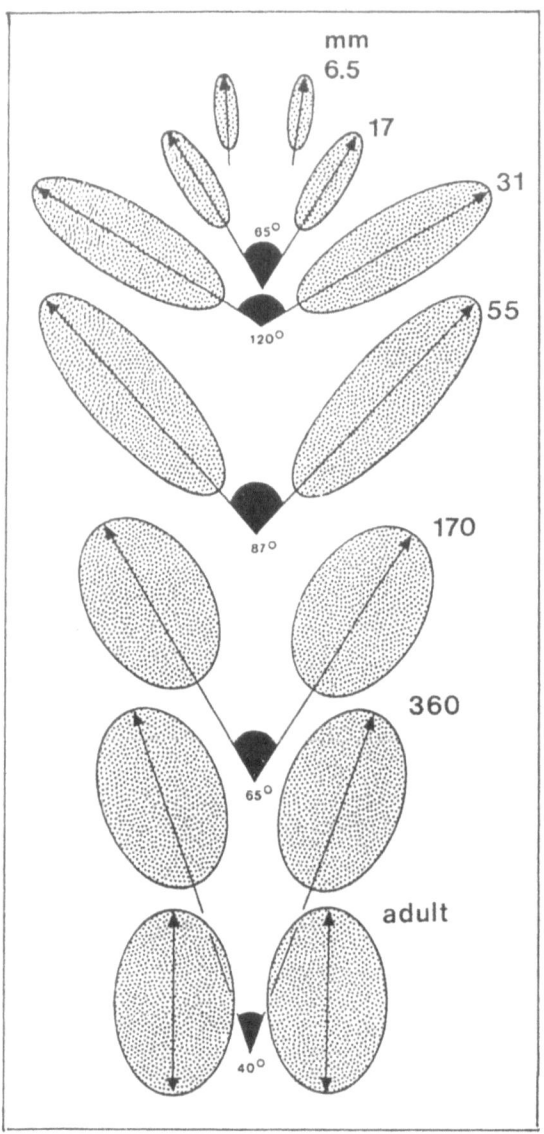

Fig. 49. Angle between axes of testes and body axes of 6.5-mm CRL embryos to adult. Note the changes during testicular descent. Ventral view

4.2
Discussion

4.2.1
Development, Structure and Function of the Gubernaculum Hunteri

The gubernaculum testis in humans was intensively described by Moskowicz (1935, 1939), Backhouse (1964, 1982) Heyns (1987) and others. Backhouse also pointed out that it was von Haller (1749) who first described the structure which Hunter (1762) named "gubernaculum testis" (cited from Backhouse 1964).

In contrast to Moskowicz (1935, 1939) and Backhouse (1964), who did not find a gubernacular anlage until the embryo is about 13.5–16 mm CRL, we saw the first appearance of the abdominal part of the gubernaculum at stage 13–14 CC (5–7 mm CRL, phase I) when the mesonephric fold comes into contact with the conus inguinalis. This mesenchymous conus is not identical with the muscular conus inguinalis as described by Klaatsch (1890), which in rodents and insectivores might pull the testis through the inguinal canal (for nomenclature in rodents see also van der Schoot 1996b).

During the eighth week of embryonic development (stages 20–23 CC, phase II), the gubernaculum penetrates the abdominal wall, turns medially and ends subcutaneously near the anlage of symphysis pubis. Referring to Moskowicz (1935) therefore, we distinguish three parts of the gubernaculum: abdominal, interstitial and subcutaneous.

According to Wensing (1988), the outgrowth of the extra-abdominal part of the gubernaculum brings about the intra-abdominal migration of both the testis and gubernaculum. In the early stages this might be possible only indirectly since the abdominal gubernaculum is connected with nothing other than the genital ducts proximally. However, the intra-abdominal part remains relatively short compared with the interstitial part and pelvic growth. Thus the cranial insertion of the gubernaculum comes relatively closer to the internal inguinal ring. Outgrowth and elongation of the gubernaculum are assumed to be due to cell division and increase of extracellular matrix (Wensing and Colenbrander 1986). But the mechanism of gubernaculum guiding is not understood. We propose that the outgrowing genital branch of the genitofemoral nerve which is found as early as in phase II in the vicinity of the gubernaculum directs also the outgrowth of the latter as a sort of guide rail.

Most diagrams in embryology textbooks [e.g. Crouch 1965, Patten 1968 (modified after Hertwig), Wartenberg 1990, Drews 1993, Sinowatz 1999] show the gubernaculum since its first appearance extending from the caudal pole of the testis to the bottom of the scrotum. In contrast, in early stages we found neither direct contact between the testis and gubernaculum – this is in line with Bramann (1884); see also for discussion Youssef and Raslan (1971) – nor an extension of the gubernaculum to the bottom of the scrotum. In early developmental stages the gubernaculum inserts into the mesenchyme of the genital ducts. It does not run dorsally behind this mesenchyme to the caudal pole of gonads as Hayek (1969) postulated. Before the passage of the testis into the scrotum, the gubernaculum inserts distally into the pubic region. Skworzoff (1924) assumed this insertion to be the fixed point which allows forces to affect the testicular descent which might be generated by involution of gubernaculum with simultaneous increase of the foetus body length.

With the start of the foetal period, the gubernaculum increases in length and volume. Likewise, the gonads grow considerably. Thus, in 70–100-mm embryos (phase III and IIIa), the caudal pole of the testis overgrows the genital ducts and for the first time comes into direct contact with the now swollen gubernaculum. This observation contradicts those of Hadziselimovic and Kruslin (1979), and Hadziselimovic and Herzog (1990), who did not find a direct contact between testes and gubernaculum during descent, only a connection between cauda epididymidis and gubernaculum. These authors therefore agree with Holstein (1969), that it is the epididymis which is the driving force in testicular descent.

The sliding of the testis over both ducts is a crucial event in testicular descent, comparable with Caesar's passing the Rubikon. Henceforward the anlagen of the epididymis and ductus deferens have their final position, dorsolateral to the testis. This gliding process is made possible by the regression of the müllerian duct. Non-regression of the müllerian duct due to MIS deficiency may cause testicular maldescent (see for discussion Hutson et al. 1990, 1997; Lane and Donahoe 1998; Lyet et al. 1996). Whether MIS is also involved in gubernacular swelling (Hutson et al. 1994; Schwindt et al 1997) remains obscure.

In females, the müllerian ducts are already fused in the midline, and the ligamentum latum has formed a barrier which hinders a further descent of the ovary (Moskowicz 1935).

From its origin in the second month up to the seventh month, when testicular descent starts, the gubernaculum progressively increases in diameter and length. We measured lengths and diameters of gubernacula in embryos from 22 mm CRL to foetuses of 36 cm CRL. Whereas the increase in length is proportional to that of the embryo, the diameter increases by a factor of 55 that is about 3.5 times more than the growth of the embryo. Comparable measurements were made by Eberth (1904) and Youssef and Raslan (1971). The latter started with 48-mm embryos and measured the three parts of the gubernaculum separately, but the relatively great length of the scrotal gubernaculum in 75-mm CRL embryos is conspicuous. In sum, his measurements show that at the beginning of the seventh month, the gubernaculum definitely shortens and becomes thicker. The increase in diameter is accompanied by the change of structure from a mesenchymal to a jelly-like tissue.

It is generally accepted that this swelling reaction of the gubernaculum (Phase IIIa–V) – due to increase in glycosaminoglycans (Heyns et al. 1990) – is a prerequisite for the passage of the testis through the inguinal channel into the scrotum (see e.g. Hadziselimovic 1983; Hutson 1986; Heyns and Hutson 1995; Husman and Levy 1995). Recent studies suggest a direct androgen stimulation of the human gubernaculum (Hosie et al. 1999) as previously shown in animals, e.g. Elder et al. (1982). The hormonal control of testicular descent has been extensively discussed by Hutson et al. (1990). For more details the reader is also referred to Hutson et al. (1990, 1997).

New findings also demonstrate that *Insl3* from the insulin-like superfamily may regulate growth and differentiation of the gubernaculum in an androgen-independent way (Zimmermann et al. 1999; Nef and Parada 1999). Since *Insl3* is specifically expressed in Leydig cells its product may be identical with the postulated factor from the testis called descendin (Fentener van Vlissingen et al. 1988; Wensing 1988;

Husman and Levy 1995; Visser and Heyns 1995) involved in rapid proliferation of the gubernaculum. The regression of the so-called cranial suspensory ligament (Lee and Hutson 1999) is found to be independent of *Insl3* (Zimmermann et al. 1999; for nomenclature see also Sect. 3).

Whether the function of gubernaculum in testicular descent is only passive by dilating the inguinal canal or whether an active function by traction occurs has been controversially discussed. The traction theory either by the cremaster muscle or by contractile components of the gubernaculum proper is reviewed by Heyns and Hutson (1995), who pointed out that besides the contractile properties of the gubernaculum its firm attachment to the testis and scrotum is a condition sine qua non for the pulling of the testis into the scrotum.

Whereas in phase II the gubernaculum consists of cell-rich undifferentiated mesenchyme, phase III gubernacula of both sexes contain a network of smooth muscle cells mainly in the abdominal and inguinal part of gubernacula. This confirms findings of Youssef and Raslan (1971) that in 75-mm CRL human embryos, smooth muscle cells start to differentiate from mesenchymal cells and contradicts Heyns's (1987) observations on human foetuses that neither striated nor unstriated muscle fibres can be discerned in the gubernaculum. The smooth muscle cells may cause shortening of the cranial part of the gubernaculum not for descent into the scrotum but rather for positioning the testes over the inguinal ring (Hutson et al. 1997). In our opinion, smooth muscle fibres have no part in drawing the testicles into the scrotum as hypothesised by Youssef and Raslan (1971), since no firm attachment of the gubernaculum at the scrotum exists during descent as shown by our own preparation and also by Gier and Marion (1970) and Heyns (1987) in human foetuses.

Striated musculature is found at the periphery of the gubernaculum as part of the strip-like cremaster muscle. Single muscle fibres are also found in the core of the abdominal and the interstitial gubernaculum at phase IV. Seemingly, muscle fibres detach from the abdominal musculature not only to run caudally, forming the cremaster muscle, but also bending cranially and medially within the gubernaculum. Yet, this musculature is not as conspicuous as the cremaster sac in rodents and pigs (Wensing 1986, 1988) and its significance in human testicular descent is questioned. However, van der Schoot (1996a,c) reconsiders a similar development and function of gubernacula in placental mammals with muscle cells as constitutive element. Contraction of striated musculature of the gubernaculum was considered by Curling 1840 (cited from Heyns and Hutson 1995) as a main force in testicular descent not only in rodents but also in humans. In our opinion, the effect of striated musculature on shortening and migration of the gubernaculum can only be an additional force since most of the fibrils are oriented transversally. Beside these circular fibrils, only a few bundles and some single fibrils are found running longitudinally to the axis of gubernaculum. Skworzoff (1924) also argues that the role of the few muscle fibres within the gubernaculum is not in an active contraction but in a simple stabilisation of the gubernaculum.

It is questionable whether the striated muscle bundles found in the rat gubernacular bulb differentiate from the outer mesenchyme of the gubernaculum (Radhakrishnan et al. 1979; Zimmermann et al. 1999). With the exception of some eye muscles, all skeletal muscles are derivatives of the somites (Christ et al. 1990). Thus, the layers of striated muscles and even the isolated fibres of skeletal musculature found in the gubernaculum must have been split off from the anlagen of abdominal muscles.

In phase Va, when the testis has passed the inguinal canal and glides into the scrotum, the shrinkage of the gubernaculum is evident. This was also demonstrated by Heyns (1987) who measured a decrease in wet:dry mass ratio after descent. But it is unlikely that this shortening of the gubernaculum pulls the testis into the scrotum (see for discussion Klaatsch 1890; Heyns and Hutson 1995). Bramann (1884) already pointed out that shrinkage of the gubernaculum cannot pull the testis because no insertion of the gubernaculum into the bottom of the scrotum exists. But the shrinkage of the gubernaculum as well as connective tissue within the scrotum allows the entrance of the testis and epididymis into scrotum. In humans, this is not an active process. Interestingly, according to Youssef and Raslan (1971), the critical weight for the testis to become scrotal was 120 mg. Below this weight the testes remained abdominal. These results are rather an argument against the traction theory. The idea of the weight of the testis initiating the descensus is a very old hypothesis. But in 1848, Hyrtl held against it that the foetus is standing head down in the uterus.

We agree with Gier and Marion (1970), Heyns (1987) and others, that there is no evidence of a connection between the gubernaculum and the scrotal fold. Therefore, we reconstructed the 170-mm foetus to show the thin subcutaneous gubernaculum moving in the pubic region. While the testis passes the inguinal channel, this connection is dissolved, and the gubernaculum follows the genital branch of the genitofemoral nerve into the scrotum. Thus, the loosening from this abutment may be a prerequisite for the last part of descent. After inguinal passage, the gubernaculum is bolt-like and is only fixed proximally. During the migration of the testis into the scrotum this shrinkage goes on and forms two thin ligaments which come from the testis and the epididymis and merge distally.

Intra-abdominal pressure is another force taken into consideration as either a main or marginal power in moving the testis with its annexes through the inguinal canal (see for discussion Bramann 1884; Weil 1884; Moszkowicz 1935 "Physiologischer Gleitbruch"; Attah and Hutson 1993; Heyns and Hutson 1995). This idea is supported by Kaplan et al. (1986), who found human infants having abdominal wall defects combined with cryptorchidism "prune belly syndrome" (see for discussion Hutson and Beasley 1992). A further mechanism of human testicular descent favoured by Klaatsch (1890) is the eversion of the gubernaculum, a mechanism typical for rodents.

An effect of the genitofemoral nerve on gubernacular migration was implied from transsection experiments in animals after which the testis fail to descend (Beasley and Hutson 1988). The neurotransmitter calcitonin gene-related peptide (CGRP) was assumed to play a role in this concert since this peptide causes contraction of cultured rodent gubernacula (see for discussion Beasley and Hutson 1987; Griffiths et al 1993; Hutson and Beasley 1992; Hutson et al. 1990; Shono et al. 1995; Terada et al 1994a,b). In a recent paper from Hutson's laboratory (Schwindt et al. 1999), the conclusion was drawn that CGRP from the genitofemoral nerve (GFN) may affect gubernacular migration by release from the sensory nerves (as part of the cremaster reflex), rather than from motor nerves as previously thought. That a rhythmic twitching of the gubernaculum may direct the distal leading tip of the testis (Hutson et al. 1994) has yet to be proven in human foetuses.

Finally, Blechschmidt (1955) considered the role of the gubernaculum in testicular descent only as passive within morphogenetic movements occurring between the growing chondrogenic pelvis and the integument of the lower abdominal wall. A constant length of the connective tissue strands including the gubernaculum in this process was postulated.

The remnants of the gubernaculum in the adult male is the faint ligamentum scrotale testis at the reflection zone of the tunica vaginalis.

We can conclude that the main function of the gubernaculum in testicular descent proper, the inguinal passage (that means penetration of the abdominal wall layers) and migration into the scrotum, is the widening of the inguinal channel and the positioning of the testis and epididymis in front of the inner inguinal ring. Thus, it makes important arrangements for the penetration of the abdominal wall.

Shrinkage of the whole gubernacular mass and detaching of the distal point of fixation are prerequisites for the further migration of the testis into the scrotum by help of intra-abdominal pressure. The gubernaculum and the testis slide into the loose scrotal mesenchyme since there is no obstacle in this way.

Cases of suprapubic ectopic testis as attested by Felix (1911) and Campell (in Hasche-Klünder 1961), thus, might be due to the continued adherence of the gubernaculum from the symphyseal region.

Hormonal factors and perhaps growth factors, e.g. EGF (Cain et al. 1994), may be directly responsible for the sex-specific difference in growth and structure of the gubernaculum. The functional significance of genes like *Insl3* (Nef and Parada 1999; Zimmermann et al. 1999) and *Hoxa10* (Rijli et al. 1995; Satokata et al. 1995) has yet to be elucidated in human. The latter gene was found by Kolon et al. (1999) to be altered in boys with cryptorchidism.

4.2.2
Development and Significance of Processus Vaginalis Peritonei

The origin and relations of the processus vaginalis peritonei was very controversially discussed by the first investigators (see for discussion Bramann 1884). Today, it is generally accepted that this is a structure common to all species with testicular descent though differences are obvious. As has been emphasised by Gier and Marion (1970), in rodents and rabbits no processus vaginalis forms until days after birth due to a different pattern of testis descent. In humans and other mammals, however, this blind pouch of the abdominal cavity develops very early during ontogenesis. To the best of our knowledge, Bardeen (1907) was one of the first to show schematically the three-dimensional arrangement of the processus vaginalis in a human embryo at the end of the embryonic period. But the relationships to the gubernaculum are not drawn by the author. We show the appearance of the processus vaginalis in phase II from stage 20 CC onwards. The firm attachment of the peritoneum to the abdominal part of the gubernaculum gives rise to an excavation when the interstitial gubernaculum penetrates the abdominal wall. Thus, the interstitial gubernaculum assumes a retroperitoneal position. The floor of the processus vaginalis is found at the ventromedial side of the interstitial gubernaculum pointing to the subcutaneous gubernaculum. Thus, direction and extent of the elongating vaginal process are in accordance with the outgrowth of the gubernaculum.

In contrast to our observations, Hadziselimovic and Herzog (1990) stated that the processus vaginalis was visible for the first time in 91-mm CRL foetuses. Backhouse (1964) found only a slow growth of the processus vaginalis and the cremaster muscle until shortly before descent. He believes that the growing cremaster brings about the processus vaginalis to invade the gubernacular mesenchyme. Gier and Marion (1970), moreover, propose that the inguinal ring must have expanded sufficiently by the swelling gubernaculum to let the abdominal fluids pass into the processus. The increased abdominal pressure is supposed to be transferred to the processus vaginalis and to induce its elongation toward the scrotum. Moszkowicz (1935) indicated sex differences in the extension of the processus vaginalis, which were no longer visible in 50-mm CRL female foetuses (diverticulum Nucki). This is in contrast to our observations and those of Mitchell (1938/39), that the development of the processus vaginalis is similar in both sexes.

According to our results, the firm attachment of the peritoneum to the gubernaculum is the precondition for the formation of the blind sac of the processus vaginalis peritonei. Its continuous distal enlargement coincides with the growth of the gubernaculum which is always "a nose length" ahead.

Broman (1921) also observed this early development of the processus vaginalis which arose just at the adhesion point of the gubernacular anlage at the abdominal wall. Thus at phase IV, shortly before inguinal descent, both, gubernaculum and processus vaginalis extend distally from the external inguinal ring. Therefore, the outgrowth of the vaginal process is independent of the position of testis.

The results of Bergh et al. (1978) in rat embryos might also argue for the dependence of the processus vaginalis on the gubernaculum since the development of the processus vaginalis stopped after distal gubernaculotomy. But caution is necessary when conclusions are drawn from experiments made in rodents as already pointed out above. In these animals, the cremaster muscle forms a bilaminar sac in the outer layer of processus vaginalis (Wensing 1988), and causes the extension of the latter. Thus, according to Shono et al. (1999), enervation of the cremasteric muscles may prevent the extension of the processus vaginalis and induce a relative ascent of the testis as the rat grows.

Clarnette et al. (1996), however, proposed two phases of development of the processus vaginalis in the rat, first the mechanism described above, second, the dependence on active gubernaculum migration which seems to be essential for its extension into the scrotum.

A hitherto unresolved question is the modality of migration of the gubernaculum into the floor of the scrotum since its lower end is free and unattached to the scrotal mesenchyme (see Sect. 4.2.1). Hutson and Beasley (1992), alternatively to the traction theory, propose that enzymatic digestion might clear a space ahead of the gubernaculum. Reduction in the turgidity of the gubernaculum may be a prerequisite for the testis achieving its final position (Levy and Husmann 1995). Thus, the processus vaginalis prolongates distally until it reaches the bottom of its hemiscrotum.

The cranial part of the vaginal process usually becomes occluded around birth (vestige of vaginal process) but in some cases it is already obliterated before birth (Mitchell 1938/39). From clinical studies, Momoh (1982) postulated three mechanisms of fibrous obliteration of the processus vaginalis. Finally, Barthold and Redman (1996) found an association of epididymal anomalies with patent processus vaginalis which indicates that both, epididymal differentiation and closure of the vaginal process may be androgen dependent.

4.2.3
Role of the Scrotum in Testicular Descent

The testosterone derivative dihydrotestosterone is considered to control the development of the penis and scrotum starting in embryos at about 45 mm CRL (Wilson et al. 1981). The scrotal anlage prior to testicular descent contains two chambers filled with loose mesenchyme separated by a septum consisting of connective tissue and blood vessels. Before the entrance of the testis, musculature is only present in the form of smooth muscle fibres in the forming tunica dartos.

The development of the unpaired scrotum from the paired genital swellings and the formation of the septum seems to occur not by fusion but by local differentiation processes (Starck 1982). Different positions of the scrotum exist in Mammalia (Gier and Marion 1970) with the postpenial one typical of humans. From the elevation of the penis, Moszkowicz (1935) deduced that most of scrotal tissue derives from an area originally cranial to the penis and only the caudal part of the scrotum comes from genital swellings. Posterior shifting and unification of genital swellings is terminated at 10–12 weeks (Hinrichsen's results in Wartenberg 1990).

Doubtlessly, the descent of the testis into the scrotum allows a lower temperature for the testis which is necessary for normal spermatogenesis (Setchell 1998). The descended scrotal testis is the phylogenetic primitive condition as shown by Werdelin and Nilsonne (1999) and four hypothesis concerning the evolution of testis are reviewed by the latter: 1, the scrotum provides a lower temperature than in the core body; 2, the scrotum evolved as a signalling device in social competition; 3, the epididymis needs a cold storage for sperm; and 4, the scrotum is a training ground for having few but high-quality sperms. These authors also pointed out, that since testicular descent is very costly and dangerous, other mechanism are developed in testicond animals to solve the problem of temperature sensitivity.

The significance of the scrotal anlage for a complete testicular descent was emphasised by van der Schoot et al. (1995): In freemartins, the gubernaculum and the processus vaginalis develop in a male-like pattern. Yet, due to absence of androgen influence, no scrotum is formed and the expansion of the processus vaginalis beyond the pubic region is inhibited. Likewise, with excision of scrotal skin, cryptorchid testis can be produced in rats (Shono and Suita 1995). On the other hand, the passage of the testis into the scrotum seems to be growth-dependent: Testicular atrophy by estradiol treatment causes testicular maldescent also in rats (Spencer et al. 1993).

4.2.4
Position and Role of Annexe of the Testis During Descent

Controversy exists concerning the role of the annexes, mainly the epididymis, in testicular descent (see also for discussion Hutson et al. 1990; Heyns and Hutson 1995). Holstein (1969), Hadziselimovic and Kruslin (1979), as well as Hadziselimovic and Herzog (1990) observed the cauda epididymis to descent first into the scrotum. According to the latter authors, the intraperitoneal testis is embraced during descent by the considerably elongated and curled ductus epididymis. Immediately before and during the inguinal passage, the rapidly growing head of the epididymis is supposed to press the testes caudad. The significant role of the epididymis in testicular descent was also postulated by Mininberg (1987) and Mininberg and Schlossberg (1983) referring to patients with cryptorchidism. But whether epididymal malformations are really the cause of testicular maldescent or if both are effects of androgen failure (McMahon et al. 1995) remains to be determined. However, clinical data of Elder (1992) argue against a significant role of the epididymis in testicular descent.

Usually, embryology textbooks do not demonstrate how the annexes come into their final position. According to our knowledge, the only one who paid attention to this process was Moszkowicz (1935) showing the free inferior pole of the testis to be in front of the genital ducts. We found that in phase III and IIIa of the descent, the testes slide over the genital ducts and come into an anterior position. From that stage on, the annexe lies in its definitive position to the testis. In the meantime, the müllerian duct regresses and the wolffian duct and the mesonephros differentiate into the epididymis. At phase IV the testis and cauda epididymis have deeply dived into the jelly gubernaculum with the epididymis seemingly slightly more distally at first. However, during the inguinal passage (phase V), it is the testis which has the leading position. The role of the epididymis as the prime mover (Holstein 1969; Hadziseli-movic and Kruslin 1979; Hadziselimovic and Herzog 1990) is in our opinion not tenable. The epididymis extends only minimally farther distal than the caudal pole of the testis and this is without functional significance.

Interestingly, in species with natural testicondy, the tail of the epididymis is advanced in the vaginal process and comes to lie closer to the integument (Bedford 1978). Those species do not develop a scrotal anlage. Thus, whether the testis or the epididymis has the leading position depends on the special mechanical prerequisites. We suppose that a large scrotum containing loose mesenchyme is able to receive both the testis and the adhering epididymis. But that epididymal traction pulls the testis into the scrotum without a fixation point outside its body is as impossible as Münchhausen pulling himself by his own hair out off the marsh.

During testicular descent not only the epididymis but also the deferens duct and the supplying blood vessels are intimately associated with the testis. Adequate elongation of these structures has to occur during descent. Resistance to elongation of theses structures may be partly responsible for the delay of the testis to immediately enter the scrotum after inguinal passage (Gier and Marion 1970).

Sampaio et al. (1999) showed that the human foetal testis is always supplied by a testicular and a deferential artery but additional arteries with varying origin are commonly found. As shown in Sect. 3, the testicular artery develops from one of the most caudal mesonephric arteries whereas the others regress (Mall 1910; Felix 1911; Blechschmidt 1973; Gasser 1975; Hill 1907, with comparable results in pig embryos).

5 Origin of the Caudal Ligaments of the Ovary and Uterus

Most embryology and anatomy textbooks (e.g. Williams and Warwick 1980; Wartenberg 1994) hold the view that the ovarian ligament is a derivative of the gubernaculum since the genital ducts divide the gubernaculum into an upper and lower part. This

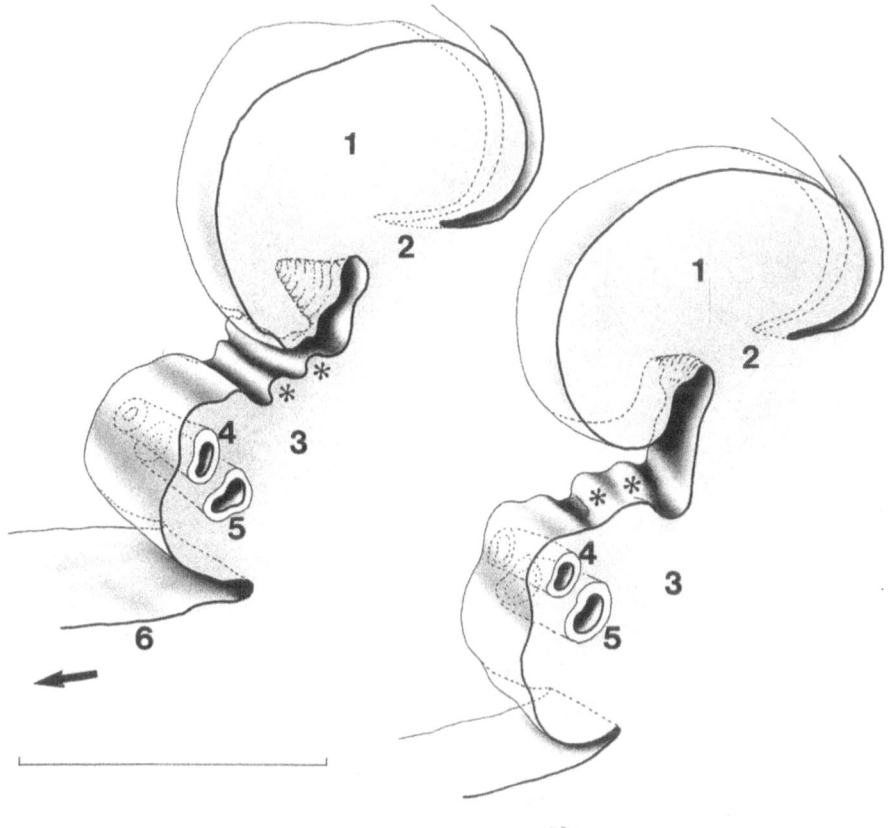

Fig. 50. Bipartite three-dimensional reconstruction of ovary, genital ducts and anlage of the duplication of ovarian ligament of an embryo of 21 mm CRL, from medial to lateral right side. *1*, ovary; *2*, mesovarium; *3*, mesonephros; *4*, müllerian duct; *5*, wolffian duct; *6*, abdominal wall. *Asterisks* show the double anlage of ovarian ligament, the *big arrow* points ventrally. *Bar*, 0.5 mm

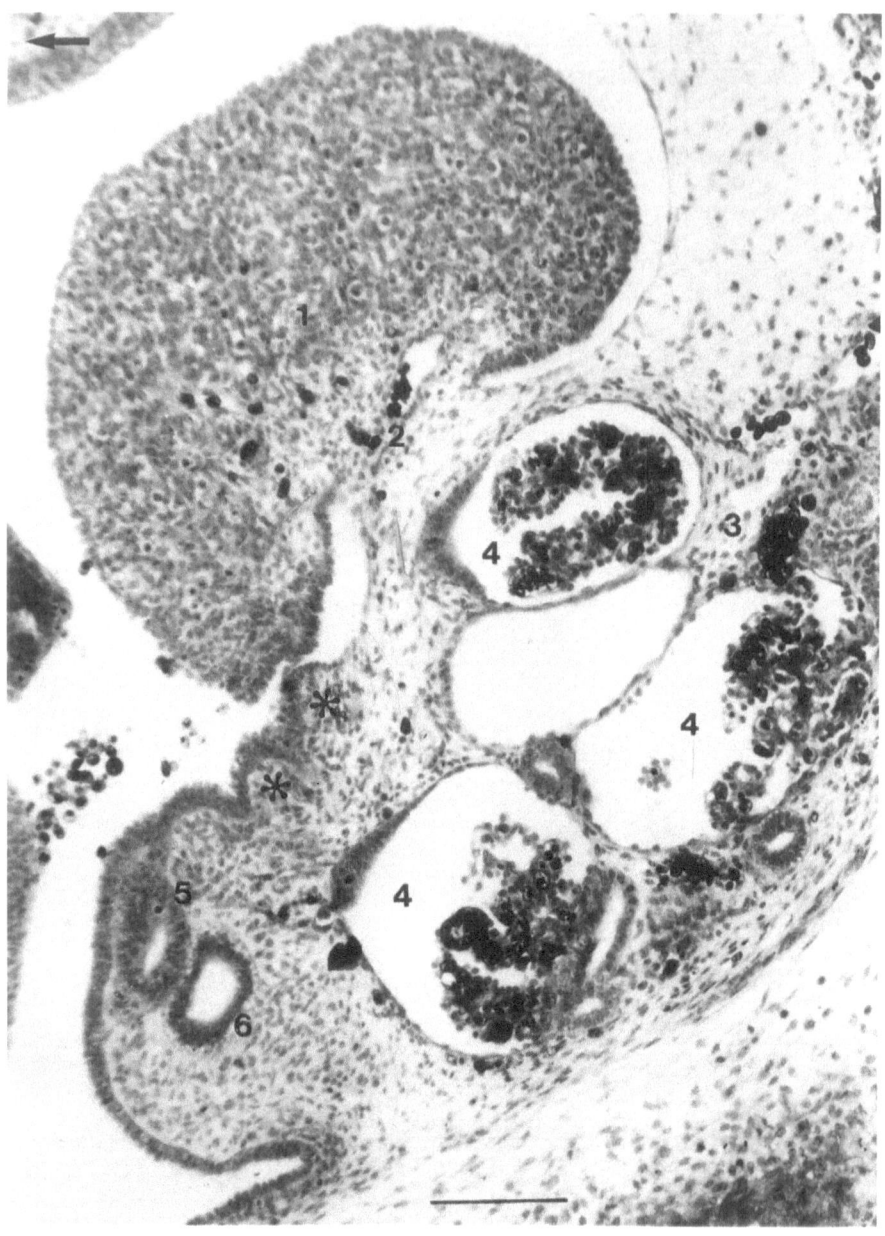

Fig. 51. Histological sagittal section through the 21-mm CRL female embryo reconstructed in Fig. 50. Same view. *1*, ovary; *2*, mesovarium; *3*, mesonephros; *4*, glomeruli of mesonephros; *5*, müllerian duct; *6*, wolffian duct. *Asterisks* show the double anlage of ovarian ligament at the presumptive dorsal superior angle of the uterus. *Bar,* 0.1 mm

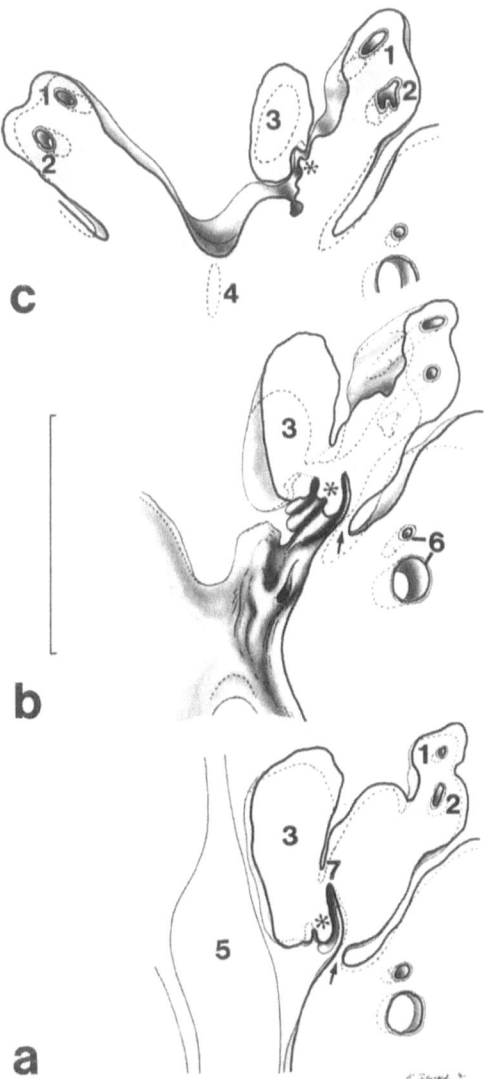

Fig. 52a–c. Partial reconstruction of a 29-mm CRL female embryo with the anlage of the ovarian ligament at the presumptive dorsal superior angle of the uterus, from dorsal (**a**) to ventral (**c**) side. *1*, müllerian duct; *2*, wolffian duct; *3*, ovary; *4*, fused müllerian ducts with anlage of uterus (luminary); *5*, contour lines of rectum; *6*, right inferior epigastric vessels; *7*, mesovarium. The *arrow* shows urogenital mesentery; the *asterisk* shows dorsal superior angle of uterus, connection of uterus and ovary, and the anlage of ovarian ligament. *Bar*, 1 mm

conception is necessary as long as the assumption that the gubernaculum is connected with the lower pole of the gonad from its beginning is not questioned. The diagrams are often based on the concept of Hertwig (1907), showing the gubernacu-

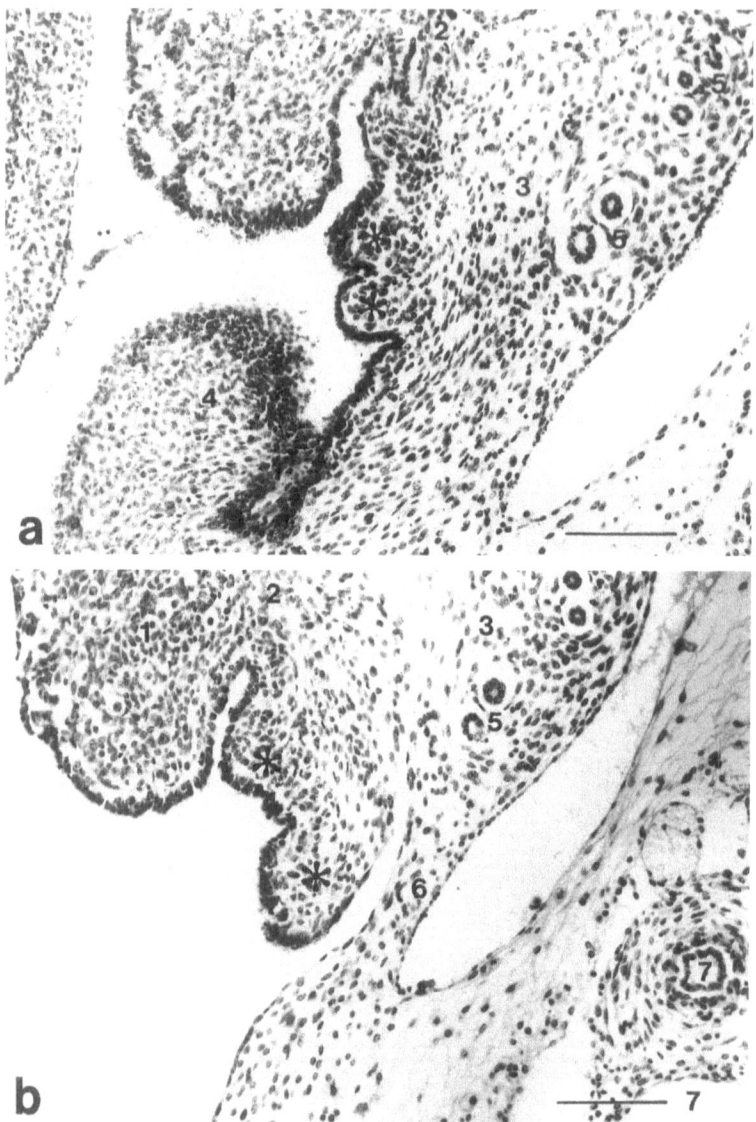

Fig. 53a–c. Sagittal section of the 29-mm CRL embryo reconstructed in Fig. 52. Dorsal view right side. **a** Anlage of ovarian ligament as demonstrated in Fig. 52c. **b** Further dorsally (compare Fig. 52b). *1*, ovary; *2*, mesovarium; *3*, mesonephros; *4*, anlage of uterus, dorsal superior angle of uterus; *5*, glomeruli of mesonephros; *6*, urogenital mesentery; *7*, right inferior epigastric vessels. *Asterisks* indicate the double anlage of ovarian ligament. *Bars,* 0.1 mm

lum as single strand extending from the gonads to the bottom of the scrotum (Patten 1968; see for discussion van der Schoot 1996b). In contrast, Frankl (1904, cited after Neumann 1933), Moskowicz (1935) and van der Schoot (1996b) emphasise a different

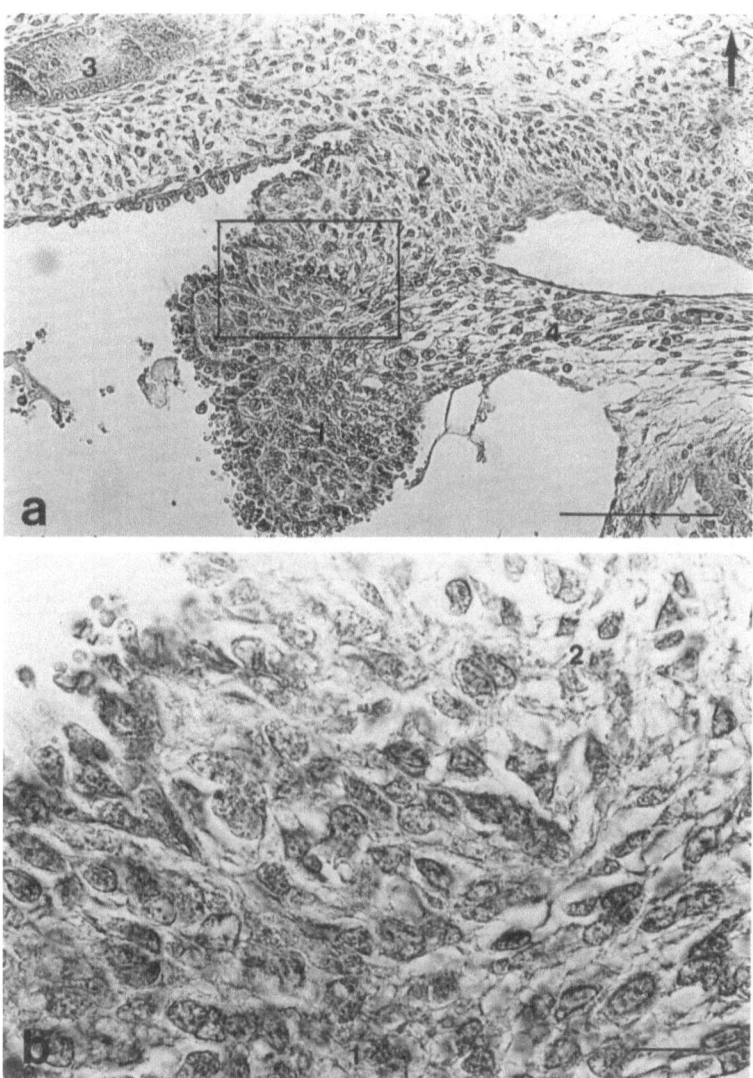

Fig. 54a,b. Transverse section through a 38-mm CRL female embryo to show the right dorsal superior angle of uterus anlage. **a** Overall view of this region with part of the ovary. **b** Detail of *rectangle* in a of the boundary between ovary and anlage of ovarian ligament (punctuated in *rectangle*). *1*, extremity of ovary; *2*, anlage of ovarian ligament; *3*, müllerian duct; *4*, mesovarium. *Bars*, 0.1 mm

origin of these structures, which is in better agreement with our data on gubernaculum development. With graphic reconstructions, histological sections and scanning electron microscopy we will give striking evidence for different origins of the ovarian ligament, round ligament and mesovarium.

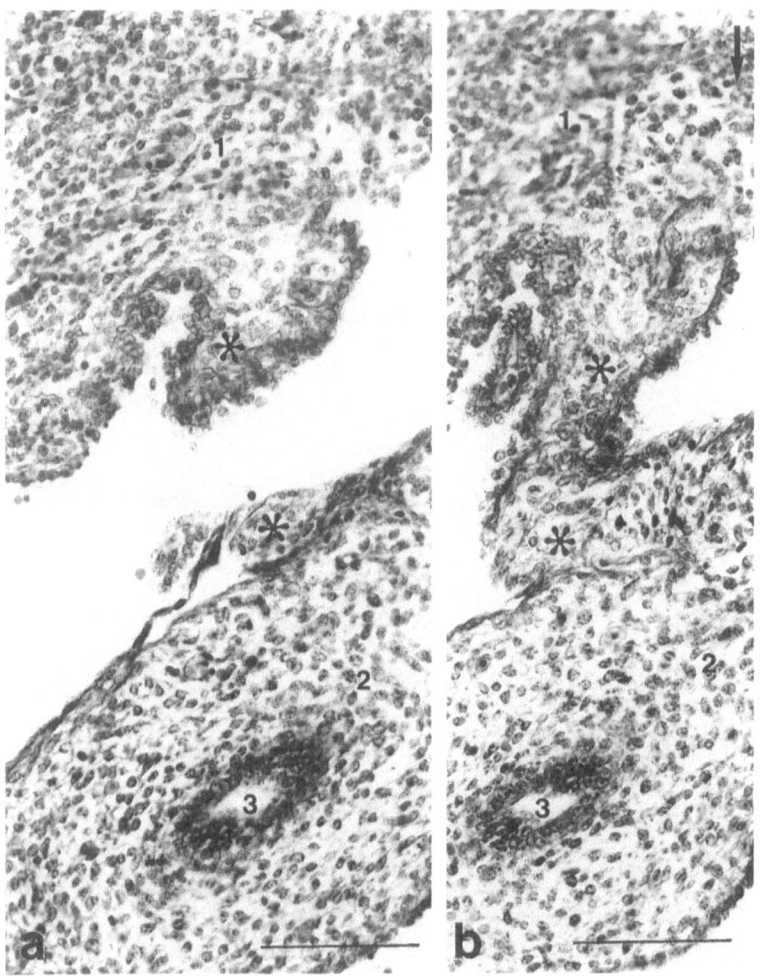

Fig. 55a,b. Transverse sections through the dorsal superior angle of the uterus anlage of a CRL foetus <50 mm. **a** Left anlage of ovarian ligament (*asterisks*). **b** The subsequent section with connection between ovary and anlage of uterine tube or mesonephros. *1*, ovary; *2*, urogenital fold; *3* müllerian duct. *Bars*, 0.1 mm

The ovarian ligament is first visible in an embryo of 21 mm CRL (stage 20 CC; Figs. 50, 51). It appears as a double, U-shaped fold elevating from the dorsal mesenchyme of both genital ducts at their crossing-over and extends to the caudal pole of the ovary. It is not identical with the mesovarium. As has been shown in Figs. 13 and 15, the abdominal gubernaculum develops on the opposite, ventral side of the genital ducts. Thus, the loose mesenchyme of the latter separates ovarian ligament and abdominal gubernaculum. It should be noted that the gubernaculum has developed earlier than the ovarian ligament.

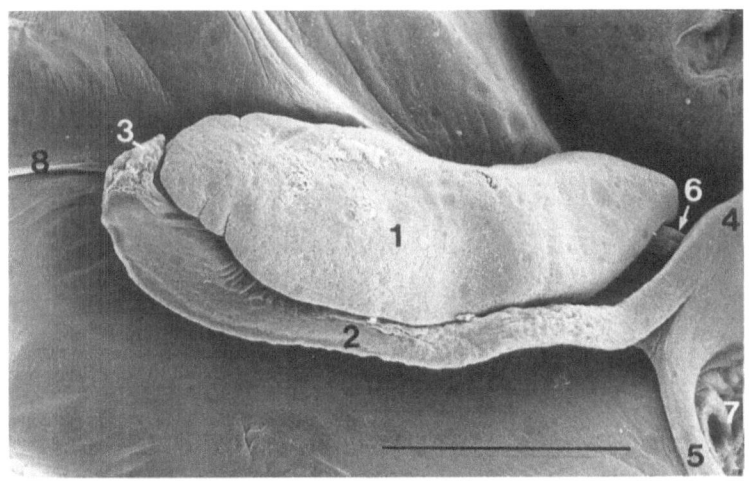

Fig. 56. SEM micrograph of the right ovary of a 78-mm CRL foetus. Lateral view. *1*, ovary; *2*, genital duct/uterine tube; *3*, fimbria of uterine tube anlage; *4*, anlage of uterine fundus; *5*, abdominal part of gubernaculum/round ligament; *6*, ovarian ligament; *7*, orifice of ureter into urinary bladder; *8*, cranial mesonephric ligament. *Bar,* 1 mm

The double-folded anlage of the ovarian ligament is more pronounced in 29-mm CRL embryos. The reconstruction (Fig. 52) clearly reveals its relationship to the mesovarium and the urogenital mesentery: The ovarian ligament anlage extends from the dorsal superior angle (Tubenwinkel) of the uterus anlage towards the ovary (Figs. 52, 53).

In the 38-mm CRL embryo, the mesenchymous ovarian ligament is embedded in the dorsal mesenchyme of the genital ducts at their crossing-over [future superior angle of the uterus (Tubenwinkel); Fig. 54]. The relations are unchanged.

A transverse section through a 50-mm embryo (Fig. 29a) exhibits the insertion of the thick abdominal gubernaculum at the presumptive ventral side of the superior angle of the uterus (Tubenwinkel). Both structures are separated by the mesenchyme of the genital ducts and no continuity is visible. In the close-up picture (Fig. 55), the origin of the ovarian ligament from a double peritoneal fold can be suspected.

Likewise, the SEM micrograph of a 78-mm CRL foetus (Fig. 56), with definitive relations of the ovary, uterine tube and uterus anlage demonstrates the discontinuity of the round ligament anlage and the ovarian ligament.

The duplication of the ovarian ligament also becomes clear in the partial reconstruction of a 90-mm CRL embryo (Fig. 57). Again, its original development from a U-shaped peritoneal fold is evident.

Interestingly, in male embryos a broad peritoneal connection between the testis and dorsal mesenchyme of the wolffian and the regressing müllerian duct exists before the testis has glided over both ducts (Fig. 30). However, this strong fold is more likely a part of the mesentery of the testis than a separate peritoneal structure.

The histological section of the 90-mm foetus (Fig. 58) reveals structural differences between the dorsal and ventral branches of the ovarian ligament. The dorsal part

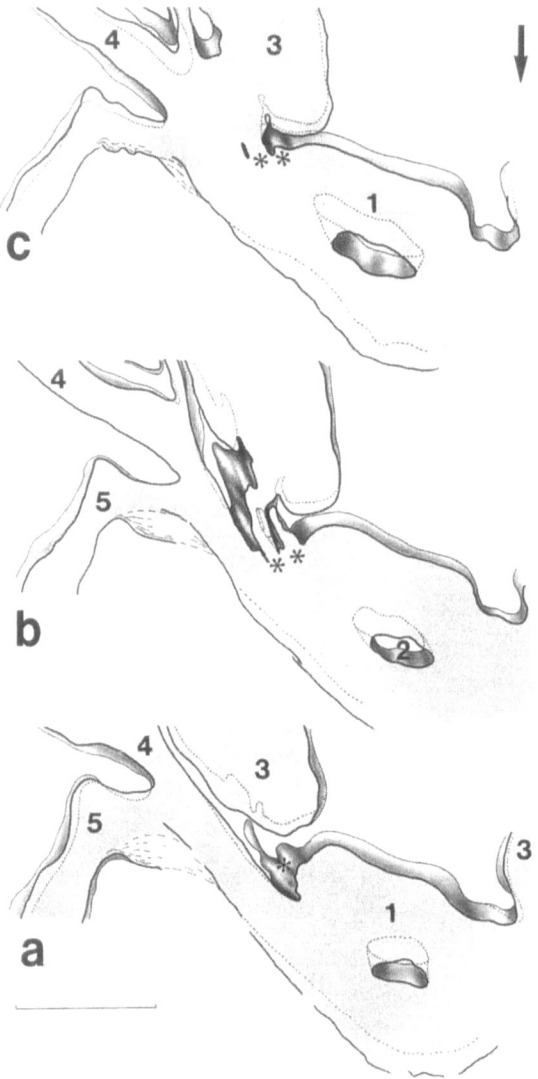

Fig. 57a–c. Tripartite reconstruction of the left superior angle of the uterus of a 90-mm foetus from caudal (**a**) to cranial (**c**) end. *1*, uterus; *2*, uterine cavity; *3*, ovary; *4*, uterine tube; *5*, abdominal part of gubernaculum/round ligament. *Asterisks* show double anlage of ovarian ligament. *Bar*, 1 mm

appears richer in cells with smooth muscle cells immigrating from the myometrial anlage. The reconstruction (Fig. 59a) shows the relations between the mesovarium, round ligament and ovarian ligament. The latter now contains the ovarian branch of the uterine artery. The differences in structure between the ovarian ligament on one hand and of the mesovarium and the round ligament on the other are obvious (Fig. 59b).

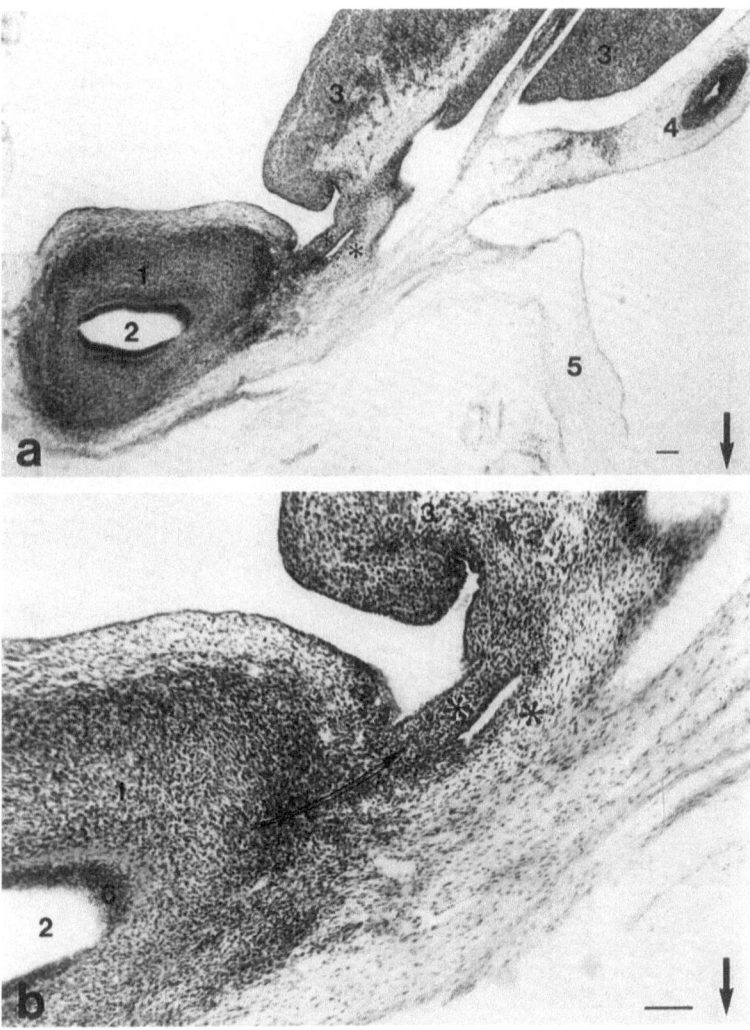

Fig. 58a,b. Transverse section through the same foetus as in Fig. 57. **a** Overall view with inner genital organs and ligaments. **b** Detail of the dorsal superior angle of the uterus with the anlage of ovarian ligament. Note the emigrating smooth muscle cells (*arrow*). *1*, myometrium; *2* uterine cavity; *3*, ovary; *4*, part of developing uterine tube; *5*, abdominal part of gubernaculum/round ligament; *6*, endometrium. *Asterisks* show the duplication of ovarian ligament. *Bars*, 0.1 mm

The synopsis (Fig. 60) demonstrates the differences in the development of the lower gonadal ligaments and the gubernaculum during male and female differentiation referring to stages of testicular descent.

The most impressive peculiarity is the swelling and shortening of the gubernaculum in males whereas in females the gubernaculum persists as a long and thin ligament. The latter forms the round ligament of the uterus. In males, the gonad dives

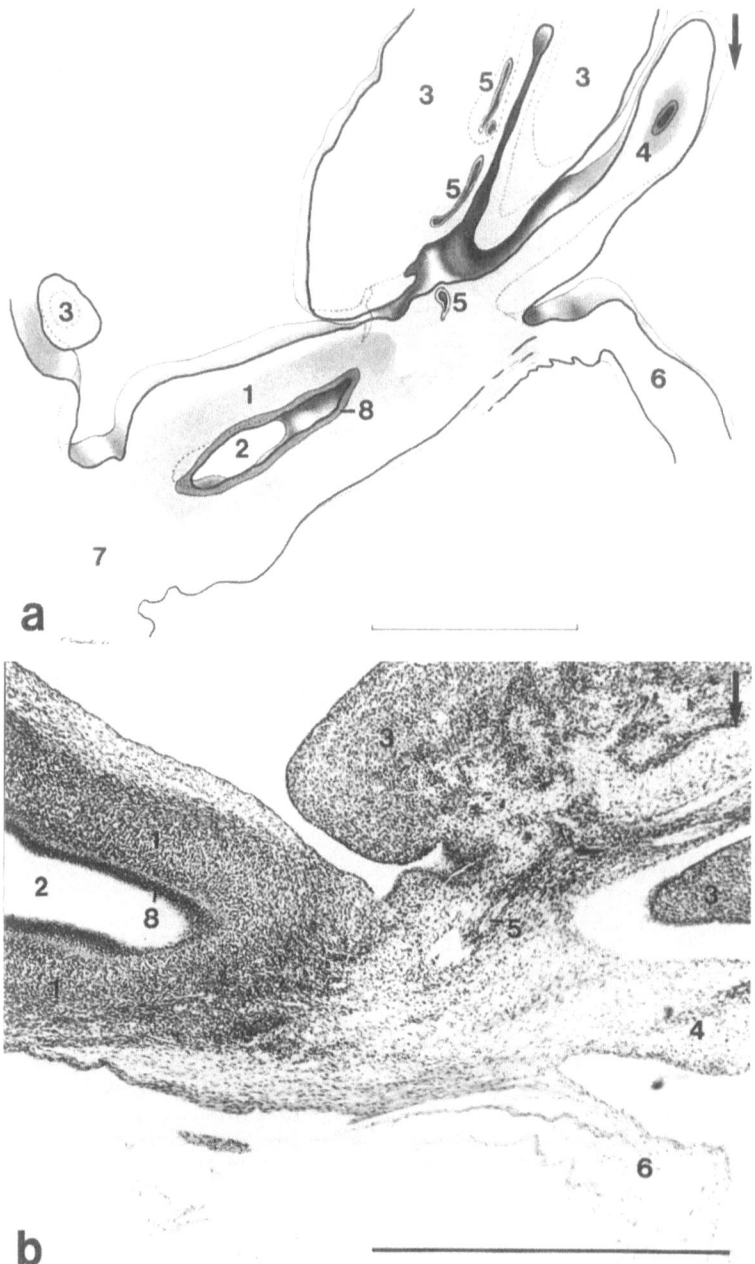

Fig. 59. a Partial reconstruction of the inner genital organs of the 90-mm CRL female foetus as shown in Fig. 58, but with a more cranial view of the ovarian branch of the uterine artery. **b** Histological section of the same region. *1*, uterus/myometrium; *2*, uterine cavity; *3*, ovary; *4*, anlage of uterine tube; *5*, ovarian branch of uterine artery; *6*, abdominal part of gubernaculum/ round ligament; *7* broad ligament; *8*, endometrium. *Bars*, 1 mm

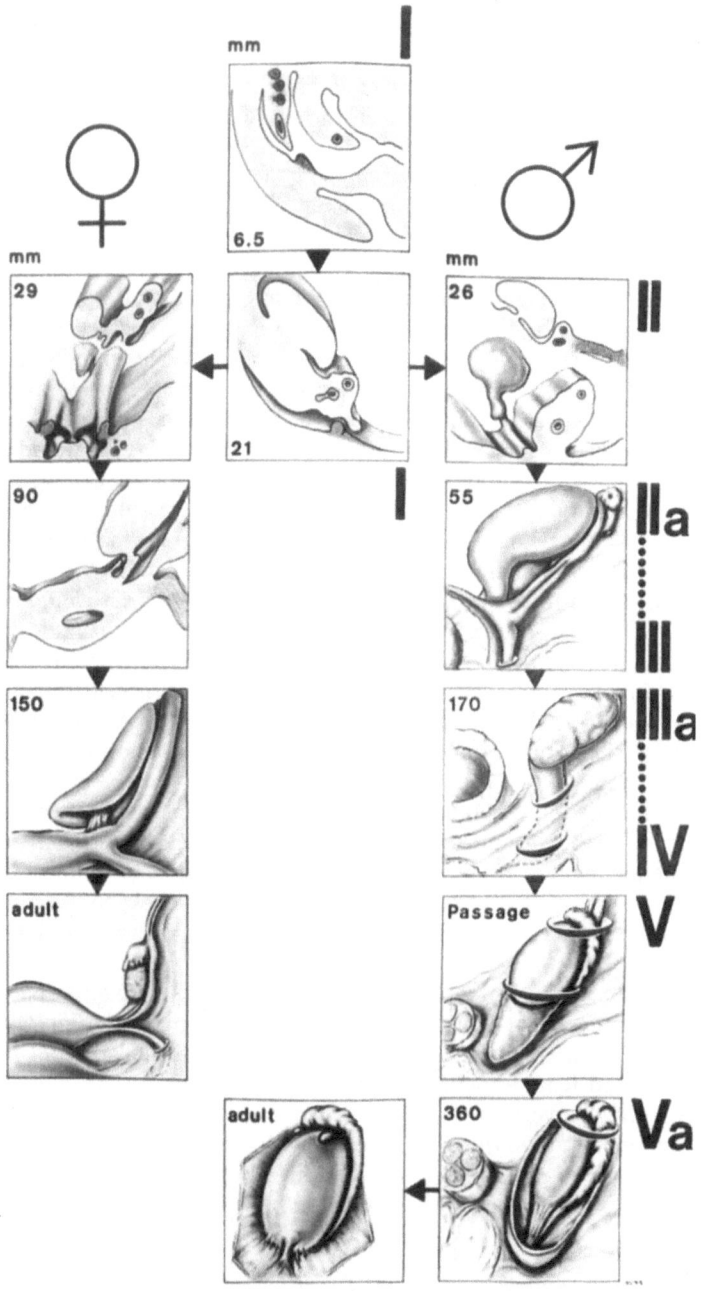

Fig. 60. Synopsis of the development of the ovarian ligament (*left*) compared with the phases of testicular descent (*right*). Note the early indifferent stages in the middle

into the gubernaculum while the ligamentous connection with the mesenchyme of the genital ducts disappears.

In females, the gonad never comes into contact with the gubernaculum. But a new structure, the ovarian ligament, forms and connects the ovary with the dorsal superior angle (Tubenwinkel) of the uterus anlage.

According to Ludwig (1993), a caudal gonadal ligament is formed in stage 23 CC from the caudal pole of the gonads and proliferates in the space of the two genital ducts. In females, the caudal gonadal ligament (anlage of ovarian ligament) is supposed to fuse with the inguinal ligament (anlage of round ligament of uterus) and both together form the gubernaculum. We, however, found the anlage of the ovarian ligament to arise independently of the dorsal mesenchyme of the genital ducts at their crossing. The abdominal gubernaculum has been formed earlier, as shown in Sect. 4, and inserts at the ventral side of the later superior angle of the uterus (Tubenwinkel).

Investigation of the histological sections and the three-dimensional graphic reconstructions reveal that no fusion of the ligaments occurs. They are always separated by the thick mesenchymal layer of the genital ducts or the müllerian duct. Thus, we cannot agree with Clara (1967), Drews (1993), Grosser and Ortmann (1966), Horstmann and Stegner (1966), Hutson et al. (1996) and many others who considered the ovarian ligament as the cranial and the round ligament as the caudal part of the female gubernaculum. We are rather in line with Frankl (1904, cited after Neumann 1933), Moskovicz (1935), and van der Schoot (1996b), who proposed that the two structures have a different origin. According to van der Schoot (1996b), the ovarian ligament, "is the remnant of the peritoneal covering of the mesonephric ventromedial surface, which is the side where the gonads emerge." Yet, we regard the ligament as a proliferation from the peritoneum of the urogenital fold and thus as a new formation containing the ovarian branch of the uterine artery.

Whereas Horstmann and Stegner (1966), assume a nearly fibrous structure of the ovarian ligament up to birth, we found smooth musculature emigrating from the myometrium into the ligament in 90-mm CRL foetuses. The double-stranded ligament contains the anlage of the bipartite smooth musculus tensor ovarii. Its functional significance in the sampling movements of the uterine tube is emphasised by Horstmann and Stegner (1966).

According to our results, the round ligament is analogous to the male gubernaculum and, therefore, with the scrotal ligament of the testis. Immunostaining with anti-smooth muscle actin has revealed a network of smooth musculature in the gubernaculum of male and female foetuses. In females, these smooth muscle cells reveal no continuity with the myometrium of the developing uterus proximally. Thus, we agree with van der Schoot (1996b) that there is no evidence to support the notion that the musculature of the round ligament derives directly from the myometrium anlage.

6 Summary and Conclusions

Testicular descent has to be divided into the turn-out of the testis and epididymis from the abdomen proper and an inner abdominal descent of genital organs. Both events are closely related to and depend on the development and reorganisation of ligaments, mainly the gubernaculum Hunteri. These seemingly unambiguous events are controversially described since the first description of the gubernaculum, and results and specifics of other species were intermingled with data from humans, thus giving more confusion than lucidity in this important step of gonadal development.

Here, we concentrate on human embryos, chronologically investigated by serial sections, scanning electron microscopy, three-dimensional reconstructions, microdissection and immunohistochemistry.

The first question to be answered was whether a real inner descent of gonads occurs. We demonstrated this inner descent by showing the relations of the gonads, mesonephros, cranial mesonephric ligament and the anlage of the diaphragm with the vertebral segments. No explosion-like increase in certain vertebral segments was observed which might simulate a gonadal descent.

The inner descent is coupled with the growth of the gonad (ovary or testis), the involution of mesonephros, the descending septum transversum or the anlage of the diaphragm, and the intercalated cranial mesonephric ligament. This ligament always inserts medially at the border between the gonad and mesonephros in close relationship to the abdominal ostium of the müllerian duct, a region where hydatides often occur.

In contrast to the testes, the ovaries arrive very early – 20–25 mm CRL – at their definitive position of S2/3 (level of linea terminalis pelvis), yet, are transversally oriented. The cranial gonadal ligament does not exhibit notable increase in length during inner descent. It does not contain blood vessels. While regressing in both sexes, it will be replaced by the plica formed by the ovarian vessels, that is the suspensorium ligament of the ovary as known in adults.

The second point to be investigated was the origin, development, structure and fate of the gubernaculum Hunteri as well as the processus vaginalis peritonei. Their arrangement and composition is crucial for testicular descent. We discriminated five phases of their development and differentiation.

Phase I characterises the early development of the gubernaculum of stage 14 CC (5–7 mm CRL) embryos. It arises as conus inguinalis and connects the abdominal wall lateral to the umbilical artery with the caudal part of the mesonephric fold. It is in this early stage that the localisation of the inner inguinal ring is defined.

In phase II, stage 20–23 CC (21–30 mm CRL), three parts of the gubernaculum – abdominal, interstitial and subcutaneous – can be distinguished. The processus vaginalis peritonei appears with its dorsal layer firmly adhering to the ventral side of the gubernaculum. The gubernaculum inserts cranially into the mesenchyme of the genital ducts at their crossing-over. Opposite to it, but at some distance, a ligament connects the caudal pole of the testis with the dorsal mesenchyme of the genital ducts. In female embryos, the anlage of the ovarian ligament appears as a U-shaped, double peritoneal fold.

Phase II is subdivided in phase IIa (32–55 mm CRL), characterised by an enormous increase in length and volume of the gubernaculum and also an enlargement of the processus vaginalis peritonei.

In phase III, sex-specific differences in gonadal position and gubernacular structure can be observed for the first time. Testes increase in volume and come close to the mesenchyme of the genital ducts. The caudal pole of the testis overlaps both ducts.

We also subdivide this phase into phase IIIa (about 100 mm CRL) where two very important events occur in male foetuses: 1, the swelling of the gubernaculum, and 2, the gliding of the testis across the genital ducts. This gliding is permitted by both, the regression of the müllerian duct and the involution of the cranial mesonephric ligament.

During phase IV (about 170 mm), the testis and epididymis immerse into the gubernaculum which is now as broad as the testis. With the interstitial gubernaculum pushing forward, the inner inguinal ring becomes distinct. The caudal poles of the testis as well of the epididymis are positioned in front of the internal inguinal ring.

Smooth muscle cells are found within the gubernaculum, and striated muscle fibres discharged from the cremaster muscle. These muscles seem to help in positioning of the testis.

The most distal, subcutaneous part of the gubernaculum inserts into the symphyseal region. The external inguinal ring develops by further advance of the interstitial gubernaculum. Likewise, the annulus inguinalis profundus widens and the testis with its annexe enters the inguinal channel.

Shortly after this, the testis passes the external inguinal ring, too, (phase V, 7-month foetus, about 25 cm CRL.). The distal fixation of the subcutaneous gubernaculum loosens and now the bolt-like gubernaculum is freely suspended from the caudal pole of the testis and epididymis.

During this passage, the gubernaculum follows the genital branch of the genitofemoral nerve, accompanying and preceding the gubernaculum into the scrotum.

During and after the inguinal passage and entrance into the scrotum (phase Va, about 36 cm), the gubernaculum shrinks and persists merely as thin strands running from the caudal pole of the testis and epididymis which fuse distally to a single structure, the scrotal ligament of adults.

Thus, we conclude that during no phase of testicular descent does the gubernaculum Hunteri connect the testis with the bottom of the scrotum. Therefore, the testis cannot be pulled into scrotum by contraction of the gubernaculum. According to our observations in humans, the gubernaculum has the following functions and qualities:

1. In early developmental stages, the thin and inconspicuous gubernaculum connects the mesenchyme of genital ducts proximally with the symphyseal region distally.
2. Firm attachment to the peritoneal layer and distal growth of the gubernaculum permits herniation of the processus vaginalis peritonei.
3. The gubernaculum thickens mainly proximally with its greatest diameter in the seventh month. Thus it widens the inguinal channel.

4. The testis has also increased in volume and is positioned in front of the internal inguinal ring with help from the gubernaculum.
5. The passage of the testis through the inguinal channel is caused by shrinkage of the gubernaculum and intra-abdominal pressure.
6. The detachment of the fixing-point of the gubernaculum in the symphyseal region and further shrinkage permits the migration of the testis into the scrotum.

The processus vaginalis peritonei arises with the outgrowth of the gubernaculum to which one side of its wall is firmly connected. During penetration of the abdominal wall and diving into the loose mesenchyme of the scrotum of both structures, the processus vaginalis peritonei only holds a gliding-promoter function .

As generally accepted, abdominal pressure seems to be an important factor in herniation into scrotum.

Growth and differentiation of the gubernaculum was recently shown to be regulated by androgen and Insl3. The latter seems to be identical with the hitherto unknown testicular factor named descendin.

Whether MIS has a direct or only an indirect effect in testicular descent has to be clarified, and the functional role of genes like HOXA10 must be analysed.

The human scrotum, which finally receives the testis and its annexe, is already formed in early foetal stages. During the inguinal passage of the testis, neither gubernaculous material nor smooth musculature is found in the scrotal chambers. Four theories for evolution of the scrotum and testicular descent have been discussed by Werdelin and Nilsonne (1999). Clinical investigations concerning cryptorchidism and infertility (Elder 1988; Hadziselimovic and Herzog 1990), support the temperature hypothesis.

The female gubernaculum at first develops in a similar manner to the male. But it shows no swelling and no shrinkage. It persists as a round ligament of the uterus which penetrates the abdominal wall together with the processus vaginalis peritonei (rudimentary, diverticulum Nucki). The round ligament remains always fixed distally at the pubic region.

The ovarian ligament is not a part of the female gubernaculum but develops from a double, U-shaped peritoneal fold at the dorsal superior angle (Tubenwinkel) of the uterus anlage. During foetal life it already contains smooth musculature emigrating from the uterus anlage. The female gubernaculum also contains smooth muscle cells developing from the mesenchyme.

References

Attah AA, Hutson JM (1993) The role of the intra-abdominal pressure in cryptorchidism. J Urol 150: 994–996

Backhouse KM (1964) The gubernaculum testis Hunteri: Testicular descent and maldescent. Ann roy Coll Surgns Engl 35: 15–35

Backhouse KM (1982) Development and descent of the testis. Eur J Pediatr 139: 249–252

Bardeen CR (1907) Development and variation of the nerves and the musculature of the inferior extremity and the neighboring regions of the trunk in man. Amer J Anat, Vol. 6

Bardeen CR (1910) Die Entwicklung des Skeletts und des Bindegewebes. In: Keibel F und Mall FP (Hrsg.) Handbuch der Entwicklungsgeschichte des Menschen, Bd. 1, 11. Kap. S. Hirzel, Leipzig, pp 296–456

Barteczko K and Jacob M (1999a) Zur Frage des inneren Descensus testis beim Menschen unter Berücksichtigung des Zwerchfells (Septum transversum) und der sogenannten kranialen Keimdrüsenbänder. Verh Anat Ges 94 (Anat Anz, Suppl 181) p 80

Barteczko K and Jacob M (1999b) Comparative study of shape, course, and disintegration of the rostral notochord in some vertebrates, especially humans. Anat Embryol 200: 345–366

Barteczko K, Jacob M and Jacob HJ (1998) Zur Entstehung, Formveränderung und zum Schicksal des Gubernaculum Hunteri beim Descensus testis des Menschen. Verh Anat Ges 93, (Anat Anz, Suppl 180) pp 78–79

Barthold JS, Redman JF (1996) Association of epididymal anomalies with patent processus vaginalis in hernia, hydrocele and cryptorchidism. J Urol 156: 2054–2060

Baumans V, Dijkstra G, Wensing CJ (1982) The effect of orchidectomy on gubernacular outgrowth and regression in the dog. Int J Androl 5: 387–400

Baumans V, Dijkstra G, Wensing CJ (1983) The role of a non-androgenic testicular factor in the process of testicular descent in the dog. Int J Androl 6: 541–552

Baumans V, Dieleman SJ, Wouterse HS, Tol L van, Dijkstra G, Wensing CJ (1985) Testosterone secretion during gubernacular development and testicular descent in the dog. J Reprod Fertil 73: 21–25

Beasley SW, Hutson JM (1987) Effect of division of genitofemoral nerve on testicular descent in the rat. Aust N Z J Surg 57: 49–51

Beasley SW, Hutson JM (1988) The role of the gubernaculum in testicular descent. J Urol 140: 1191–1193

Bedford JM (1978) Anatomical evidence for the epididymis as the prime mover in the evolution of the scrotum. Amer J Anat 152: 483–508

Bergh A, Helander HF and Wahlquist L (1978) Studies on factors governing testicular descent in the rat – particularly the role of gubernaculum testis. Int J Androl 1: 342–356

Bjerklund Johansen TE (1988) Non-union of testis and epididymis. Description of an experimental model in the rat. Effect on testicular descent. Scand J Urol Nephrol 22: 165–170

Blechschmidt E (1955) Wachstumsfaktoren des Descensus testis. Z Anat Entwickl·Gesch 118: 175–182

Blechschmidt E (1960) Hrsg. Die vorgeburtlichen Entwicklungsstadien des Menschen. Karger, Basel, London, New York, pp 601–607

Blechschmidt E (1973) Hrsg. Die pränatalen Organsysteme des Menschen. Hippokrates Verlag, Stuttgart

Bramann F (1884) Beitrag zur Lehre von dem Descensus testiculorum und dem Gubernaculum Hunteri des Menschen. Arch Anat Physiol 310-340

Broman I (1921) Hrsg. Entwicklung der Geschlechtsorgane, Grundriß der Entwicklungsgeschichte des Menschen. Bergmann, München, Wiesbaden, pp 182-208

Cain MP, Kramer SA, Tindall DJ, Husmann DA (1994) Alterations in maternal epidermal growth factor (EGF) effect testicular descent and epididymal development. Urology 43 (3): 375-378

Cain MP, Kramer SA, Tindall DJ, Husmann DA (1995) Flutamide-induced cryptorchidism in the rat is associated with altered gubernacular morphology. Urology 46: 553-558

Christ B, Brand-Saberi B, Jacob HJ, Jacob M, Seifert R (1990) Principles of early muscle development. In: Le Douarin N, Dieterlen-Lievre F, Smith J (eds) The avian model in developmental biology: From organism to genes. CNRS, Paris, pp 139-151

Clara M (1967) Hrsg. Entwicklungsgeschichte des Menschen, B. Die Entwicklung der Keimdrüsen und ihrer Ausführungsgänge. Quelle & Meyer, Heidelberg, pp 350-372

Clarnette TD, Hutson JM (1996) The genitofemoral nerve may link testicular inguinoscrotal descent with congenital inguinal hernia. Aust N Z J Surg 66: 612-617

Clarnette TD, Hutson JM (1997) Exogenous calcitonin gene-related peptide can induce the testis to cross the scrotal septum. Br J Urol 79: 623-627

Clarnette TD, Hutson JM (1999) Exogenous calcitonin gene-related peptide can change the direction of gubernacular migration in the mutant trans-scrotal rat. J Pediatr Surg 34: 1208-1212

Clarnette TD, Hutson JM and Beasley SW (1996) Factors affecting the development of the processus vaginalis in the rat. J Urol 156: 1463-1466

Clarnette TD, Sugita Y, Hutson JM (1997) Genital anomalies in human and in animal models reveal the mechanism and hormones governing testicular descent. B J Urol 79: 99-112

Colenbrander B, van Straaten HW, Wensing CJ (1978) Gonatrophic hormones and testicular descent. Arch Androl 1: 131-137

Crouch JE (1965) ed. Reproductive System. In: Functional Human Anatomy. Lea & Febiger, Philadelphia, pp 447-479

Dalton AJ (1955) A chrome-osmium fixative for electron microscopy. Anat Rec 121: 281

Drews U (1993) Hrsg. Taschenatlas der Embryologie, 8 Urogenitalsystem. Thieme, Stuttgart, New York

Eberth CJ (1904) Die männlichen Geschlechtsorgane. In: v Bardeleben K (Hrsg.) Handbuch der Anatomie des Menschen, II. Teil, Abt. 2, Bd VII. Gustav Fischer, Jena, pp 279-292

Elder JS (1988) The undescent testis: hormonal and surgical management. Surg Clin North Am 68: 983-1005

Elder JS (1992) Epididymal anomalies associated with hydrocele/hernia and cryptorchidism: implications regarding testicular descent. J Urol 148: 624-626

Elder JS, Isaacs JT, Walsh PC (1982) Androgenic sensitivity of the gubernaculum testis: evidence for hormonal/mechanical interactions in testicular descent. J Urol 127: 170-176

Emmen JMA, McLuskey A, Grootegoed JA and Brinkmann AO (1998) Androgen action during male sex differentiation includes suppression of cranial suspensory ligament development. Human Reprod 13: 1271-1280

Felix W (1911) Die Entwicklung der Harn- und Geschlechtsorgane. In: Keibel F und Mall FP (Hrsg.) Handbuch der Entwicklungsgeschichte des Menschen. Bd. II, 19. Kap. S Hirzel, Leipzig, pp 732-955

Fentener van Vlissingen JM, van Zoelen EJ, Ursem PJ, Wensing CJ (1988) In vitro model of the first phase of testicular descent: identification of a low molecular weight factor from fetal testis involved in proliferation of gubernaculum testis cells and distinct from specified polypeptide growth factors and fetal gonadal hormones. Endocrinology 123: 2868-2877

Fentener van Vlissingen JM, Koch CA, Delpech B, Wensing CJ (1989) Growth and differentiation of the gubernacum testis during testicular descent in the pig: changes in the extracellular matrix, DNA content, and hyaluronidase, beta-glucuronidase, and beta-N-acetylglycosaminidase activities. J Urol 142: 837-845

Fischel A (1929) Hrsg. Lehrbuch der Entwicklung des Menschen. Springer, Berlin, pp 633-638

Frey HL, Peng S, Rajfer J (1983) Synergy of abdominal pressure and androgens in testicular descent. Biol Reprod 29: 1233-1239

Frey HL, Rajfer J (1984) Role of the gubernaculum and intraabdominal pressure in the process of testicular descent. J Urol 131: 574-579

Fujikake N, Togashi H, Yamamoto M, Arishima K, Ueda Y, Eguchi Y (1989) Relationship between the development of the gubernaculum and the testicular descent in the rat fetus: macroscopic and light and electron microscopic observation. Nippon Juigaku Zasshi 51: 416–424

Gasser RF (1975) ed. Atlas of Human Embryos. Harper & Row, Hagerstown, Maryland

Gier HT and Marion GB (1970) Development of the Mammalian Testis. In: Johnson AD, Gomes WR and Vandemark NL (eds.) The Testis. Vol. I. Academic Press, New York and London, pp 1–43

Goh DW, Momose Y, Middlesworth W and Hutson JM (1993) The relationship among calcitonin gene-related peptide, androgens and gubernacular development in 3 animal models of cryptorchidism. J Urol 150: 575–576

Griffiths AL (1993) The tammar wallaby (Macropus eugenii) and the Sprague-Dawley rat: comparative anatomy and physiology of inguinoscrotal testicular descent. J Anat 183: 441–450

Griffiths AL, Middlesworth W, Goh DW, Hutson JM (1993) Exogenous calcitonin gene-related peptide causes gubernacular development in neonatal (Tfm) mice with complete androgen resistance. J Pediatr Surg 28: 1028–1030

Grosser O, Ortmann R (1966) Hrsg. Geschlechtsorgane, Grundriß der Entwicklungsgeschichte des Menschen. Springer, Berlin, Heidelberg, New York, pp 140–147

Habenicht UF, Neumann F (1983) Hormonal regulation of testicular descent. Advanc Anat Embryol, Vol. 81

Hadziselimovic F (1983) Embryology of testicular descent and maldescent. In: Hadzeselimovic F (ed) Cryptorchidism: Management and implications. Springer-Verlag, Berlin, pp 11–34

Hadziselimovic F and Kruslin E (1979) The role of the epididymis in descensus testis and the topographical relationship between the testis and epididymis from the sixth month of pregnancy until immediately after birth. Anat Embryol 155: 191–196

Hadziselimovic F, Herzog B (1990) Hodenerkrankungen im Kindesalter. In: Daum R, Mildenberger H und Rehbein F (Hrsg.) Bibliothek für Kinderchirurgie, Ped/Surg. Hippokrates, Stuttgart

Hamilton WJ, Boyd ID and Mossman HW (1972) eds. Human Embryology. The Williams & Wilkins Comp., Baltimore

Hasche-Klünder R (1961) Genitalmißbildungen ohne Intersexualität, In: Overzier C (Hrsg.) Die Intersexualität, Georg Thieme Verlag, Stuttgart, pp 497–515

Hayek v H (1969) Die Entwicklung der Harn- und Geschlechtsorgane. In: Alken CE, Dix VW, Goodwin WE, Wildbolz E (Hrsg.) Handbuch der Urologie, Springer, Berlin, New York, pp 1–50

Hertwig O (1907) Hrsg. Die Elemente der Entwicklungslehre des Menschen und der Wirbeltiere. Gustav Fischer, Jena

Heyns CF (1987) The gubernaculum during testicular descent in the human fetus. J Anat 153: 93–112

Heyns CF, de Klerk DP (1985) The gubernaculum during testicular descent in the pig fetus. J Urol 133: 694–699

Heyns CF, Hutson JM (1995) Historical review of theories on testicular descent. J Urol 153: 754–767

Heyns CF, Human JM, de Klerk DP (1986) Hyperplasia and hypertrophy of the gubernaculum during testicular descent in the fetus. J Urol 135: 1043–1047

Heyns CF, Human HJ and Werely CJ (1989) The collagen content of the gubernaculum during testicular descent in the pig fetus. J Anat 167: 161–166

Heyns CF, Human HJ, Werely CJ and de Klerk DP (1990) The glycosamino-glycans of the Gubernaculum during testicular descent in the fetus. J Urol 143: 612–617

Heyns CF, Tata R, Sargent NS, Habib FK, Chisholm GD (1993) Absence of 5 alpha-reductase activity in the gubernaculum during descent of the fetal pig testis. J Urol 150: 510–513

Hill EC (1907) On the cross development and vascularization of the testis. Amer J Anat 6: 439–459

Holstein AF (1969) Morphologische Studien am Nebenhoden des Menschen, In: Bargmann W und Doerr W (Hrsg.) Zwanglose Abhandlungen aus dem Gebiet der normalen und pathologischen Anatomie, Heft 20, 1–91 (Med. Fakultät Hamburg 1967, Habilitationsschrift) Thieme, Stuttgart

Horstmann E, Stegner H-E (1966) Harn- und Geschlechtsapparat. In: v Möllendorff (Hrsg.) Handbuch der mikroskopischen Anatomie des Menschen, Bd. VII/4. Springer, Berlin, Heidelberg, New York

Hullinger RL, Wensing CJ (1985) Descent of the testis in the fetal calf. A summary of the anatomy and process. Acta Anat 121: 63–68

Hosie S, Wessel L, Waag KL (1999) Could testicular descent in humans be promoted by direct androgen stimulation of the gubernaculum testis? Eur J Pediatr Surg 9: 37–41

Husmann DA, Levy JB (1995) Current concepts in the pathophysiology of testicular descent. Urology 46: 267–276

Hutson JM (1985) A biphasic model for the hormonal control of testicular descent. Lancet 24: 419–421

Hutson JM (1986) Testicular feminization: a model for testicular descent in mice and men. J Pediatr Surg 21: 195–198

Hutson JM (1994) Testicular descent: the first step towards fertility. Int J Androl 17: 281–288

Hutson JM, Beasley SW (1987) The mechanisms of testicular descent. Aust Paediatr J 23: 215–216

Hutson JM, Beasley SW (1992) eds. Descent of the Testis. Edward Arnold, London

Hutson JM, Hasthorpe S and Heyns CF (1997) Anatomical and functional aspects of testicular descent and cryptorchidism. The endocrine Society, Endocrine Review 18 (2): 259–280

Hutson JM, Williams MP, Fallat ME, Attah A (1990) Testicular descent: new insights into its hormonal control. Oxf Rev Reprod Biol 12: 1–56

Hutson JM, Terada M, Zhou B, Williams MPL (1996) Normal testicular descent and the aetiology of cryptorchidism. Advanc Anat Embryol, Vol. 132

Hutson JM, Baker M, Terada M, Zhou B and Paxton G (1994) Hormonal control of testicular descent and the cause of cryptorchidism. Reprod Fertil Dev 6: 151–156

Hyrtl J (1848) Hrsg. Handbuch der topographischen Anatomie und ihrer praktisch medizinisch-chirurgischen Anwendungen, 1.Bd., Wilhelm Braumüller, Wien

Jacob M, Barteczko K und Jacob HJ (1998) Entwickelt sich das Ligamentum ovarii proprium als eine eigenständige Struktur und somit Gubernaculum-unabhängig? Verh Anat Ges 93, (Anat Anz, Suppl 180) p 120

Jirasek JE (1983) ed. Atlas of human prenatal morphogenesis. Martinus Nijhoff Publishers, Boston

Johansen TE, Clausen OP, Nesland JM (1989) The effect of non-union of testis and epididymis and of cryptorchidism on the development of epididymis and ductus deferens in the rat. Andrologia 21: 441–448

Kaplan LM, Koyle MA, Kaplan GW, Farrer JH, Rajfer J (1986) Association between abdominal wall defects and cryptorchidism. J Urol 136: 645–647

Kassim NM, McDonald SW, Reid O, Bennett NK, Gilmore DP and Payne AP (1997) The effects of pre- and postnatal exposure to the nonsteroidal antiandrogen flutamide on testis descent and morphology in the Albino Swiss rat. J Anat 190: 577–588

Kersten W, Molenaar GJ, Emmen JA and Schoot P van der (1996) Bilateral cryptorchidism in a dog with persistent cranial testis suspensory ligaments and inverted gubernacula: report of a case with implications for understanding normal and aberrant testis descend. J Anat 189: 171–176

Klaatsch H (1890) Über den Descensus testiculorum. Morphol Jb 16: 587–646

Kollmann J (1907) ed. Handbuch der Entwicklungsgeschichte des Menschen. 2.Teil, Gustav Fischer Verlag, Jena

Kolon TF, Wiener JS, Lewitton M, Roth DR, Gonzales ET Jr. and Lamb DJ (1999) Analysis of homeobox gene Hoxa10 mutations in cryptorchidism. J Urol 161: 275–280

Lam SK, Clarnette TD, Hutson JM (1998) Does the gubernaculum migrate during inguinoscrotal testicular descent in the rat? Anat Rec 250: 159–163

Lane AH and Donahoe PK (1998) New insights into Mullerian inhibiting substance and its mechanism of action. J Endocrinol 158 (1): 1–6

Larsen WJ (1993) ed. Human Embryology. Churchill Livingstone, New York

Lee SM, Hutson JM (1999) Effect of androgens on the cranial suspensory ligament and ovarian position. Anat Rec 255: 306–315

Levy JB, Husmann DA (1995) The hormonal control of testicular descent. J Androl 16: 459–463

Ludwig KS (1993) The development of the caudal ligaments of the mesonephros and of the gonads in the human: a contribution to the development of the Gub. (H.). Anat Embryol 188: 571–577

Lyet L, Vigier B and Schoot P van der (1996) Anti-Müllerian hormone in relation to the growth and differentiation of the gubernacular primordia in mice. J Reprod Fertil 108: 281–288

Mall FP (1910) Die Entwicklung des Coeloms und des Zwerchfells. In: Keibel F und Mall FP (Hrsg.) Handbuch der Entwicklungsgeschichte des Menschen, Bd.I, 13. Kap. S Hirzel, Leipzig, pp 527–544

McMahon DR, Kramer SA, Husmann DA (1995) Antiandrogen induced cryptorchidism in the pig is associated with failed gubernacular regression and epididymal malformations. J Urol 154: 553–557

Mininberg DT (1987) The epididymis and testicular descent. Eur J Pediatr 146 Suppl 2: 28–30

Mininberg DT, Schlossberg S (1983) The role of the epididymis in testicular descent. J Urol 129: 1207–1208

Mitchell GAG (1938/39) The condition of the peritoneal vaginal processes at birth. J Anat 73: 658–661

Momoh JT (1982) Obliteration of processus vaginalis and inguinal hernial sacs in children. Can J Surg 25: 483–485

Moszkowicz L (1935) Das Gubernaculum Hunteri und seine Bedeutung für den Descensus testiculorum beim Menschen. Z Anat Entwickl Gesch 105: 37–52

Moszkowicz L (1939) Morphologie und Sinn des Descensus testiculorum. Acta neerl Morphol 2: 209–222

Nef S, Parada LF (1999) Cryptorchidism in mice mutant for Insl3. Nat Genet 22: 295–309

Netter FH (1954/65) ed. Descent of Testis, The Ciba Collection of medical Illustrations, Vol II, Reproductive Systems. Thieme Stuttgart, New York

Neumann HO (1933) Weibliche Geschlechtsorgane, 3. Die Krankheiten der Uterusbänder einschließlich Beckenbindegewebe. In: Henke F, Lubarsch O (Hrsg.) Handbuch der speziellen pathologischen Anatomie und Histologie. Springer, Berlin, pp 399–622

Noordhuizen-Stassen EN, Dijkstra G, Schamhardt HC, Wensing CJ (1983) Compensatory development of a patent vascular supply to the testis after intra-abdominal transection of its main blood vessels. Int J Androl 6: 509–519

Patten BM (1968) ed. Human Embryology, McGraw Hill, New York

Politzer G and Zeitlhofer J (1958) Die Mißbildungen der männlichen Geschlechtsorgane. In: Schwalbe E (Hrsg.) Die Morphologie der Mißbildungen des Menschen und der Tiere. III. Teil, XVIII Lieferung. G. Fischer, Jena, pp 886–904

Radhakrishnan J, Morikawa Y, Donahoe PK, Hendren WH (1979) Observations on the gubernaculum during descent of the testis. Invest Urol 16: 365–368

Rajfer J, Walsh PC (1977) Hormonal regulation of testicular descent: experimental and clinical observations. J Urol 118: 985–990

Rijli FM, Matyas R, Pellegrini M, Dierich A, Gruss P, Dolle P, and Chambon P (1995) Cryptorchidism and homeotic transformations of spinal nerves and vertebrae in Hoxa-10 mutant mice. Proc Natl Acad Sci USA 92: 8185–8189

Sadler TW (1998) Hrsg. Medizinische Embryologie. Thieme, Stuttgart, New York, pp 316–317

Sampaio FJB and Favorito LA (1998) Analysis of testicular migration during the fetal period in humans. J Urol 159: 540–542

Sampaio FJB, Favorito LA, Freitas MA, Damiao R and Gouvela E (1999) Arterial supply of the human fetal testis during its migration. J Urol 161: 1603–1605

Satokata I, Benson G and Mass R (1995) Sexually dimorphic sterility phenotypes in Hoxa10-deficient mice. Nature 374: 460–463

Schoot P van der (1992) Androgens in relation to prenatal development and postnatal inversion of the gubernacula in rats. J Reprod Fertil 95: 145–158

Schoot P van der (1993a) Doubt about the first phase of testis descent in the rat as a valid concept. Anat Embryol 187: 203–208

Schoot P van der (1993b) The name cranial ovarian suspensory ligaments in mammalian anatomy should be used only to indicate the structures derived from the foetal cranial mesonephric and gonadal ligaments. Anat Rec 237: 434–438

Schoot P van der (1996a) Human (and some other primates) uterine teres ligament represents a mammalian developmental novelty. Anat Rec 244: 402–415

Schoot P van der (1996b) Towards a rational terminology in the study of the gubernaculum testis: arguments in support of the notion that the cremasteric sac should be considered the gubernaculum in postnatal rats and other mammals. J Anat 189: 97–108

Schoot P van der, Elger W (1992) Androgen-induced prevention of the outgrowth of cranial gonadal suspensory ligaments in fetal rats. J Androl 13: 534–542

Schoot P van der and Elger W (1993) Perinatal development of the gubernacular cones in rats and rabbits: Effect of exposure to anti-androgens. Anat Rec 236: 399–407

Schoot P van der, Vigier B, Prepin J, Perchellet J-P and Gittenberger-de Groot A (1995) Development of the gubernaculum and processus vaginalis in freemartinism: Further evidence in support of a specific fetal testis hormone governing male-specific gubernacular development. Anat Rec 241: 211–224

Schwindt B, Doyle LW and Hutson JM (1997) Serum levels of Müllerian inhibiting substance in preterm and term male neonates. J Urol 158: 610–612

Schwindt B, Farmer PJ, Watts LM, Hrabovsky Z, Hutson JM (1999) Localisation of calcitonin gene-related peptide within the genitofemoral nerve in immature rats. J Pediatr Surg 34: 986–991

Setchell PB (1998) The Parkes Lecture: Heat and the testis. J Reprod Fertil 114: 179–194

Shono T, Suita S (1995) The effect of the excision of future scrotal skin on testicular descent in neonatal rats: a new experimental model of cryptorchidism. J Pediatr Surg 30: 734–738

Shono T, Ramm-Anderson S, Hutson JM (1994a) Transabdominal testicular descent is really ovarian ascent. J Urol 152: 781–784

Shono T, Ramm-Anderson S, Goh DW, Hutson JM (1994b) The effect of flutamide on testicular descent in rats examined by scanning electron microscopy. J Pediatr Surg 29: 839–844

Shono T, Goh DW, Momose Y, Hutson JM (1995) Physiological effects in vitro of calcitonin gene-related peptide on gubernacular contractility with or without denervation. J Pediatr Surg 30: 591–595

Shono T, Zakaria O, Imajima T, Suita S (1999) Does proximal genitofemoral nerve division induce testicular maldesdent or ascent in the rat ? Br J Urol Jnt 83: 323–326

Shono T, Hutson JM, Watts L, Goh DW, Momose Y, Middlesworth B, Zhou B, Ramm-Anderson S (1996) Scanning electron microscopy shows inhibited gubernacular development in relation to undescended testis in oestrogen-treated mice. Int J Androl 19: 263–270

Sinowatz F, Seitz J, Bergmann M, Petzoldt U, Fanghänel J (1999) Hrsg. Embryologie des Menschen. Deutscher Ärzte-Verlag, Köln

Siow Y, Fallat ME (1997) Testicular descent- a proposed interaction between müllerian inhibiting substance and epidermal growth factor. J Urol 158: 613–614

Skworzoff M (1924) Zur Frage über die den Descensus testiculorum bewirkenden Kräfte. Virchows Arch path Anat, 250: 636–640

Spencer JR, Vaughan ED Jr, Imperato-McGinley J (1993) Studies of the hormonal control of postnatal testicular descent in the rat. J Urol 149: 618–623

Spencer JR, Torrado T, Sanchez RS, Vaughan ED Jr, Imperato-McGinley J (1991) Effects of flutamide and finasteride on rat testicular descent. Endocrinology 129: 741–748

Starck D (1982) Hrsg. Vergleichende Anatomie der Wirbeltiere auf evolutionsbiologischer Grundlage. Bd. I-III. Springer, Berlin, Heidelberg, New York

Tayakkanonta K (1963) The gubernacular testis and its nerve supply. Aust N Z J Surg 33: 61–67

Terada M, Goh DW, Farmer PJ, Hutson JM (1994a) Ontogeny of gubernacular contraction and effect of calcitonin gene-related peptide in the mouse. J Pediatr Surg 29: 609–611

Terada M, Goh DW, Farmer PJ, Hutson JM (1994b) Calcitonin gene-related peptide receptors in the gubernaculum of normal rat and 2 models of cryptorchidism. J Urol 152: 759–762

Tran D, Picard JY, Vigier B, Berger R, Josso N (1986) Persistence of mullerian ducts in male rabbits passively immunized against bovine anti-mullerian hormone during fetal life. Dev Biol 116(1): 160–167

Tuchmann-Duplessis H, Haegel P (1972) eds. Illustrated Human Embryology, Vol. 2, Organogenesis. Masson & Cie., Paris

Visser JH, Heyns CF (1995) Proliferation of gubernaculum cells induced by a substance of low molecular mass obtained from fetal pig testis. J Urol 153: 516–520

Wagenen G van and Simpson ME (1965) eds. Embryology of the ovary and testis *Homo sapiens* and *Macaca mulatta*. Yale University Press, New Haven and London

Wartenberg H (1990) Hodendescensus und Entwicklung des Hodens im Kindesalter bis zur Pubertät, 27. Entwicklung der Genitalorgane und Bildung der Gameten, In: Hinrichsen K (Hrsg.) Humanembryologie. Springer, Berlin, Heidelberg, New York

Wartenberg H (1994) Entwicklung der Harn- und Geschlechtsorgane. In: Drenckhahn D, Zenker W (Hrsg). Benninghoff Anatomie, Bd. II, Urban & Schwarzenberg, München, Wien, Baltimore

Weil C (1884) Über den Descensus testiculorum, nebst Bemerkungen über die Entwicklung der Scheidenhäute und des Scrotums. Z Heilk 5: 225–288

Wells LJ (1954) Development of the human diaphragm and pleural sacs. Contr Embryol 35: 107–134

Wensing CJG (1986) Testicular descent in the rat and a comparison of this process in the rat with that in the pig. Anat Rec 214: 154–160

Wensing CJG (1988) The embryology of testicular descent. Horm Res 30: 144–152

Wensing CJG, Colenbrander B (1986) Normal and abnormal testicular descent. Oxf Rev Reprod Biol 8: 130–164

Werdelin L, Nilsonne A (1999) The evolution of the scrotum and testicular descent in mammals: a phylogenetic view. J Theor Biol 196(1): 61–72

Williams PL and Warwick R (1980) eds. Gray's Anatomy. Churchill Livingstone, Edinburgh, London, Melbourne, New York

Wilson JD, George FW, Griffin JE (1981) The hormonal control of sexual development. Science 211: 1278–1284

Youssef EH and Raslan NA (1971) Study for the factors which affect the descent of the testicles in man. Acta Anat 79: 422–444

Zimmermann S, Steding G, Emmen JM, Brinkmann AO, Nayernia K, Holstein AF, Engel W, Adham IM (1999) Targeted disruption of the Insl3 gene causes bilateral cryptorchidism. Mol Endocrinol 13(5): 681–691

Subject Index